Urban Sustainability

Editor-in-Chief

Ali Cheshmehzangi, School of Architecture, Design and Planning, The University of Queensland, Brisbane, QLD, Australia

The Urban Sustainability Book Series is a valuable resource for sustainability and urban-related education and research. It offers an inter-disciplinary platform covering all four areas of practice, policy, education, research, and their nexus. The publications in this series are related to critical areas of sustainability, urban studies, planning, and urban geography.

This book series aims to put together cutting-edge research findings linked to the overarching field of urban sustainability. The scope and nature of the topic are broad and interdisciplinary and bring together various associated disciplines from sustainable development, environmental sciences, urbanism, etc. With many advanced research findings in the field, there is a need to put together various discussions and contributions on specific sustainability fields, covering a good range of topics on sustainable development, sustainable urbanism, and urban sustainability. Despite the broad range of issues, we note the importance of practical and policy-oriented directions, extending the literature and directions and pathways towards achieving urban sustainability.

The series will appeal to urbanists, geographers, planners, engineers, architects, governmental authorities, policymakers, researchers of all levels, and to all of those interested in a wide-ranging overview of urban sustainability and its associated fields. The series includes monographs and edited volumes, covering a range of topics under the urban sustainability topic, which can also be used for teaching materials.

Ali Cheshmehzangi · Jian Zuo · Ayyoob Sharifi ·
Rongpeng Zhang · Abbas Ziafati Bafarasat ·
Jie Zhao
Editors

Nature-based and Climate Responsive Design for Urban Systems

Springer

Editors
Ali Cheshmehzangi
School of Architecture, Design and Planning
The University of Queensland
Brisbane, QLD, Australia

Jian Zuo
School of Civil Engineering and Construction Management
Adelaide University
Adelaide, Australia

Ayyoob Sharifi
Hiroshima University
Hiroshima, Japan

Rongpeng Zhang
Hunan University
Changsha, China

Abbas Ziafati Bafarasat
Oxford Brookes University
Oxford, UK

Jie Zhao
The University of Pennsylvania (UPenn)
Philadelphia, USA

ISSN 2731-6483 ISSN 2731-6491 (electronic)
Urban Sustainability
ISBN 978-981-92-2069-4 ISBN 978-981-92-2070-0 (eBook)
https://doi.org/10.1007/978-981-92-2070-0

This Springer imprint is published by the registered company Springer Nature Singapore Pte Ltd.
The registered company address is: 152 Beach Road, #21-01/04 Gateway East, Singapore 189721, Singapore

Dedicated to the scientists and environmental pioneers who insisted, often long before it was fashionable, that ecosystems hold the keys to resilience. This book is also dedicated to the communities who live closest to nature's forces. In small coastal towns, river settlements, tropical cities, and island communities across the world, people have long practiced forms of ecological wisdom that modern planning is only beginning to rediscover. Their lived knowledge, of tides, seasons, soils, and storms, offers lessons that no simulation model alone can fully capture.

Preface

We are not drowning. We are fighting.

—Brianna Fruean, an activist and environmental advocate from Samoa

This volume, *Nature-based and Climate Responsive Design for Urban Systems*, is the third book in the *Design Innovations and Resilience* trilogy. The conversation about resilience has reached a point where it can no longer be hypothetical, aspirational, or applied selectively. The terms of engagement have been fundamentally reshaped due to the urgency of the climate situation. What was formerly viewed as progressive is now insufficient, and what was once characterized as innovation is now an operational need. Therefore, this book is a repositioning of the design, governance, and sustainability of urban systems rather than an extension of previous talks.

This trilogy's first two volumes explored resilience from the perspectives of systemic adaptation and design innovation, highlighting the significance of incorporating environmental consciousness into architectural and urban practices. The insufficient integration of ecological systems as fundamental elements of urban planning, however, remained a crucial gap. All too frequently, nature has been viewed as an add-on, superimposed on urban settings rather than integrated into their logic. That gap is directly addressed in this volume. It makes the case that, to redefine the link between built environments and the natural systems they occupy, nature-based and climate-responsive design must shift from a peripheral to a fundamental framework.

The premise of this book is clear and uncompromising: cities can no longer rely on manufactured control as their main resilience strategy. The limitations of traditional methods have been made clear by the escalation of climate threats, such as flooding, heat extremes, coastal vulnerability, and ecosystem degradation. Planning models rooted in stability and robustness are challenged by dynamic environmental change; infrastructures optimized for efficiency are stretched by variability; and urban systems planned for predictability are increasingly faced with uncertainty. The key question is not whether current systems can be made better, but rather if they can essentially adapt to accommodate circumstances for which they were never intended.

This volume responds to that question through a focused exploration of nature-based and climate-responsive design as actionable strategies. Rather than proposing universal solutions or idealized models, it presents a diverse collection of case studies grounded in particular circumstances, such as urban villages, tiny coastal cities, growing economies, and quickly changing surroundings. These examples show that integrating ecological systems involves a set of concepts that must be understood, modified, and negotiated in many geographical, social, and environmental contexts rather than a single strategy.

This book's focus on water as a structuring component of urban systems is one of its distinguishing characteristics. In a variety of settings, water highlights the shortcomings of traditional design as well as the possibilities of different strategies. The case studies show how water can serve as a driver of spatial and ecological organization rather than a source of risk, whether leveraged through decentralized drainage, blue-green infrastructure, or coastal ecosystem integration. This focus is not incidental; it reflects the reality that water-related challenges, flooding, scarcity, and sea-level rise are among the most immediate and consequential impacts of climate change on urban environments.

Simultaneously, the book engages with inter-scalar and systemic viewpoints, going beyond site-based interventions. It acknowledges that, although crucial, local remedies are insufficient on their own. Connections across sizes, from the building and neighborhood to the city and region, are necessary to establish urban resilience. In order to do this, infrastructure must be rethought such that ecological systems like wetlands, vegetation networks, and coastal landscapes are seen as essential parts of urban functionality rather than as cosmetic or residual components.

There are some conflicts in the arguments presented in this volume. They bring up a number of critical issues that touch on governance, politics, and societal priorities in addition to design. In situations where there is a lot of pressure to develop, can cities reorganize their economic and geographical models to make room for ecological systems? Before climate consequences outrun institutional response, can planning frameworks change fast enough to incorporate nature-based approaches? Can investment models prioritize short-term, quantifiable results over the long-term, multifunctional value of ecological infrastructure? Can urban cultures change their perspective from one of control to one of cohabitation with environmental systems? This is arguably the most important question.

These inquiries imply clarity in challenge rather than ambiguity in direction. Though unequal and sometimes dispersed, the shift to nature-based and climate-responsive urban systems is already under progress. This book aims to integrate that shift into a logical framework that is supported by data and focused on execution. It presents design as an ongoing process of interaction with dynamic systems, where flexibility, integration, and long-term thinking become essential components of practice rather than as a finished product.

This book, the last in the trilogy, sharpens rather than ends the conversation. It shifts the discussion from isolated tactics to systemic change, from innovation to transformation. The message is clear: cities need to learn from nature in order to survive in the future, not as a choice.

From our six-country locations of Australia, Japan, China, the UK, the US, and the UAE.

St Lucia, QLD, Australia	Ali Cheshmehzangi
Adelaide, Australia	Jian Zuo
Hiroshima, Japan	Ayyoob Sharifi
Changsha, China	Rongpeng Zhang
Oxford, UK	Abbas Ziafati Bafarasat
Philadelphia, USA	Jie Zhao

Acknowledgements

We gratefully acknowledge the collective contributions that have shaped this book—the third volume of our work on "design innovations and resilience". We extend our sincere appreciation to all authors, co-authors, chapter contributors, and collaborators whose expertise, commitment, and creativity have advanced the discourse on climate-resilient cities and built environments. We also recognize the strong foundation laid by our earlier collaborative work, and the dedication and shared vision of contributors who continue to inspire this evolving field. Such partnerships cultivate meaningful dialogue, expand global networks, and strengthen a dynamic platform for research exchange and innovation in resilient design.

About This Book

Cities are entering a period in which climate realities are no longer abstract projections but daily design constraints. The planning, construction, and maintenance of urban environments must be fundamentally rethought in light of floodings, heatwaves, coastal erosion, water scarcity, and ecological deterioration. To address this change, *Nature-based and Climate Responsive Design for Urban Systems* investigates how ecologically based design innovations might improve urban resilience and livability while reshaping the built environment in response to climate pressures.

This book is the third of a trilogy that looks at climate-related design innovation in the built environment. This book focuses on design techniques that incorporate natural systems into urban infrastructure and spatial planning, whereas the previous volumes examine more general conceptual and strategic methods. The main idea is straightforward but effective: instead of working against natural processes, future cities must cooperate with them. Therefore, nature-based solutions, blue-green infrastructures, and climate-responsive urban design methods are positioned as crucial elements of modern urban systems rather than as supplemental environmental measures.

The book explores how ecologically based design techniques can promote community well-being, environmental restoration, and climate adaptation through a wide range of worldwide case studies. From inclusive public open space design and urban village revitalization to decentralized drainage systems, coastal protection tactics, and metropolitan-scale ecological planning, the chapters examine a variety of geographical dimensions and urban environments.

Case studies cover areas including Nepal, Bangladesh, Vanuatu, Mexico, Madagascar, and cities in the Pacific and Global South that are especially vulnerable to climate threats. These illustrations show how nature-based strategies can promote social participation and sustainable urban development while addressing issues including urban flooding, coastal vulnerability, water management difficulties, and biodiversity loss. Together, the chapters offer insights into how planning, landscape architecture, and urban design might function as transformative instruments in the climate period. They emphasize the significance of reconsidering the interaction between human settlements and natural processes, strengthening ties between infrastructure and landscape, and incorporating ecological systems into routine urban design decisions.

Most importantly, this book adds to the expanding corpus of knowledge on climate-responsive urbanism by combining interdisciplinary perspectives and practice-focused case studies. The book offers valuable lessons for researchers, practitioners, and policymakers seeking design-led pathways toward more resilient, adaptive, and environmentally integrated urban futures.

The book:

- Presents design-led climate solutions integrating nature-based systems into urban infrastructure.
- Explores international case studies from climate-vulnerable regions across the Global South and Pacific.
- Advances climate-responsive urban design through interdisciplinary research and practical examples.

Contents

Editors and Contributors

About the Editors

Ali Cheshmehzangi is a Professor and Head of the School of Architecture, Design, and Planning (ADP) at The University of Queensland. He has been in the World's top 2% field leader since 2021, recognized by Stanford University. He is among the top 20 global scholars in the urban sustainability research area. With a career spanning over two decades, he has made significant contributions to the academic and professional communities, with a focus on sustainable and environmentally conscious design. So far, Ali has published over 650 journal papers, articles, conference papers, book chapters, and reports. He also has 65 other academic books, some of which have received awards at the national, provincial, and municipal levels. He also received international awards and recognition for his research on urban resilience studies and sustainability research, as well as a Vice-Chancellor's award for his impactful contribution to higher education.

Jian Zuo is a Professor in Sustainable Construction at the School of Civil Engineering and Construction Management at Adelaide University. His main research interests are related to the sustainable built environment via both technical and managerial means. In particular, he has undertaken extensive research in construction and demolition waste recycling and management. As the lead researcher, he has been involved in a series of competitive grants in Australia such as Discovery Projects and Linkage Projects funded by the Australian Research Council as well as many other national key projects. He has long been engaged in the study of ecological towns, green buildings, and sustainable development of the construction industry. Published more than 100 papers in top journals, including more than 80 journal papers indexed by SCI and SSCI.

Ayyoob Sharifi is a Professor at the IDEC Institute, Hiroshima University. He also is a core member of the Network for Education and Research on Peace and Sustainability (NERPS). His research is mainly focused on urban sustainability, resilience, and climate change mitigation and adaptation. He has published extensively on these topics and is ranked among the world's top 1000 influential authors in all fields according to Stanford/Elsevier's Top 2% Scientist Rankings in 2024. Additionally, he received Clarivate's Highly Cited Researcher Award in 2024. Ayyoob actively contributes to global change research programs such as the Future Earth and has served as a lead author for the Sixth Assessment Report (AR6) of the Intergovernmental Panel on Climate Change (IPCC). The goal of his education and research activities is to inform actions toward building sustainable and peaceful communities.

Rongpeng Zhang is a Full Professor and dedicated Ph.D. Supervisor at the School of Architecture and Planning, Hunan University. His primary research pursuits encompass building performance modeling and simulation for intelligent buildings, alongside advanced built environment control for healthy buildings. He was a primary developer of several pivotal modules within the DOE-endorsed EnergyPlus building simulation engine, including VRF-HR, fault detection, and demand response modules. He was a primary developer of Commercial Building Energy Saver (CBES) platform, which obtained the 2019 R&D 100 Award as one of the 100 most significant new technologies in the world. Before joining Hunan University, he led the building science team and conducted a series of research on human-centric built environment control at Well Living Lab, which obtained the 2018 Edison Award. He also served as an assistant professor at Mayo Clinic, and a postdoctoral fellow in the Building Technology and Urban Systems Division at Lawrence Berkeley National Laboratory. He received his Doctorate in Building Performance and Diagnostics from School of Architecture, Carnegie Mellon University, and his B.S. and M.S. degrees in Built Environment and Facilities Engineering at Tongji University.

Abbas Ziafati Bafarasat is a Chartered Town Planner (MRTPI), Fellow of the Higher Education Academy (FHEA), and Senior Lecturer in Town Planning, School of the Built Environment, Oxford Brookes University, where he also leads the Certificate in Spatial Planning Studies. Abbas undertook his Ph.D. in Planning and Landscape at the University of Manchester, and his Post Doc at Technical University of Dortmund. His research interests involve healthy places, strategic planning, and education for sustainable development, with over 40 of his research outputs published in leading scientific platforms, mostly setting new directions (e.g., SIY Urbanism, Strategic Urban Design) and exploring novel horizons (e.g., bottom-up agenda setting) for addressing complex socio-environmental problems.

Jie Zhao is a lecturer of architecture in the Weitzman School of Design at the University of Pennsylvania, and President of P3Global, a Dubai-based sustainability, health, and smart building advisory firm. He served as Head of Delos Labs and an Executive Vice President at Delos. His research focuses on Human-building Interactions, including healthy buildings, total building performance, occupant behavior in buildings, sensing and controls in buildings, health, comfort, and productivity of building occupants, as well as building energy efficiency and sustainability. He has pub lished 20+ academic articles and serves as reviewers for 10+ top tier academic journals and conferences in the building science field. He frequently contributes to public media as a subject matter expert in healthy buildings, such as Wall Street Journal, CBS News, Forbes and Money. Jie received his Ph.D. in Building Performance and Diagnostics from Carnegie Mellon University and Master's and Bachelor's degrees from Tongji University.

All six editors have previously co-edited and published two books together on "*Designing Healthy Cities*" and "*Designing Healthy Buildings and Communities.*"

Contributors

Zaheer Allam School of Architecture, Design and Planning, The University of Queensland, Brisbane, QLD, Australia;
Institute for Sustainable Futures, University of Technology Sydney, Sydney, NSW, Australia

Abolghasem Azhdari School of Architecture, Design and Planning, The University of Queensland, Brisbane, QLD, Australia

Abbas Ziafati Bafarasat School of the Built Environment, Oxford Brookes University, Oxford, UK

Umberto Baresi School of Architecture and Built Environment, Faculty of Engineering, Queensland University of Technology, Brisbane, Australia;
Centre for Environment and Society, Queensland University of Technology, Brisbane, Australia

Ali Cheshmehzangi School of Architecture, Design and Planning, The University of Queensland, Brisbane, QLD, Australia;
School of Architecture and Planning, Hunan University, Changsha, China;
College of Architecture and Urban Planning, Tongji University, Shanghai, China

Igor Ishi Rubio Cisneros Tecnologico de Monterrey, Monterrey, Mexico

Efrit Elahi Department of Architecture, Gopalganj Science and Technology University, Gopalganj, Bangladesh

Jannatul Ferdousi Department of Architecture, Gopalganj Science and Technology University, Gopalganj, Bangladesh

Anup Ghimire Sunwal, Nepal

Sourav Ghosh Department of Architecture, Pabna University of Science and Technology, Pabna, Bangladesh

Ilija Gubic Faculty of Architecture, University of Belgrade, Belgrade, Serbia

Bao-jie He School of Architecture, Design and Planning, The University of Queensland, Brisbane, QLD, Australia;
Centre for Climate-Resilient and Low-Carbon Cities, School of Architecture and Urban Planning, Key Laboratory of New Technology for Construction of Cities in Mountain Area, Ministry of Education, Chongqing University, Chongqing, China;
CMA Key Open Laboratory of Transforming Climate Resources to Economy, Chongqing, China

Md Sahidul Islam State Key Laboratory of Regional Environment and Sustainability, School of Environment, Tsinghua University, Beijing, China

Greg Keeffe Department of School of Natural and Built Environment, Queen's University Belfast, Belfast, UK

Stefania Leontiadou School of Architecture, Aristotle University of Thessaloniki, Thessaloniki, Greece

Yanchen Liu State Key Laboratory of Regional Environment and Sustainability, School of Environment, Tsinghua University, Beijing, China

Jia-Qiang Lv State Key Laboratory of Regional Environment and Sustainability, School of Environment, Tsinghua University, Beijing, China

Alireza Mohammadi Department of Geography and Urban Planning, University of Mohaghegh Ardabili, Ardabili, Iran

Jyotir Nisha School of Architecture, Design and Planning, The University of Queensland, Brisbane, QLD, Australia

Manju Raut Tansen, Nepal

Rob Roggema Cittaideale, Wageningen, The Netherlands

Alessio Russo School of Architecture and Built Environment, Faculty of Engineering, Queensland University of Technology, Brisbane, Australia;
Centre for Environment and Society, Queensland University of Technology, Brisbane, Australia

Hassan Salehi Department of Horticultural Science, School of Agriculture, Shiraz University, Shiraz, Iran

Amirhosein Shabani Department of Urban Planning, Na.C. Islamic Azad University, Najafabad, Iran

Ayyoob Sharifi School of Architecture, Design and Planning (ADP), The University of Queensland, Brisbane, QLD, Australia;
The IDEC Institute and Network for Education and Research on Peace and Sustainability (NERPS), Hiroshima University, Hiroshima, Japan

Salman Shooshtarian School of Property, Construction and Project Management, RMIT University, Melbourne, Australia

Motahareh Soheilipoor Faculty of Architecture and Urban Planning, University of Art, Tehran, Iran

Bo Sun State Key Laboratory of Regional Environment and Sustainability, School of Environment, Tsinghua University, Beijing, China

Shima Taheri Department of Architecture, Isfahan (Khorasgan) Branch, Islamic Azad University, Isfahan, Iran

Tian Tang Independant Researcher, Sichuan and Chongqing, China

Ali Tehranifar Department of Horticultural Science and Landscape Architecture, Ferdowsi University of Mashhad, Mashhad, Iran

Tanha Tabassum Tisha Department of Architecture, Gopalganj Science and Technology University, Gopalganj, Bangladesh

Xin Wang College of Architecture and Urban Planning, Tongji University, Shanghai, China

Peter S. P. Wong School of Property, Construction and Project Management, RMIT University, Melbourne, Australia

Ali Rad Yousefnia School of Architecture, Design and Planning, The University of Queensland, Brisbane, QLD, Australia

Sourav Zaman Department of Architecture, Gopalganj Science and Technology University, Gopalganj, Bangladesh

Rongpeng Zhang School of Architecture and Planning, Hunan University, Changsha, China

Jie Zhao The Weitzman School of Design, The University of Pennsylvania, Philadelphia, PA, USA;
P3Global, Dubai, UAE

Ying Zheng Department of School of Natural and Built Environment, Queen's University Belfast, Belfast, UK

Jian Zuo School of Civil Engineering and Construction Management, Adelaide University, Adelaide, SA, Australia

Cities Can No Longer Outsmart Nature: Designing with Ecological Systems in the Climate era

Ali Cheshmehzangi, Jian Zuo, Ayyoob Sharifi, Rongpeng Zhang, Abbas Ziafati Bafarasat, and Jie Zhao

1 Rethinking Urban Systems in the Climate Age: An Overview

Long-held beliefs about control, predictability, and infrastructural superiority are no longer valid in urban systems. An engineering paradigm that views nature as a constraint that needs to be controlled, redirected, or neutralized has shaped cities for more than a century. In an effort to achieve stability, efficiency, and growth, rivers have been redirected, coastlines hardened, soils sealed, and microclimates changed [5]. Although this strategy has caused deeply ingrained vulnerabilities, it has also produced short-term rewards. Rather than creating new issues, the climate crisis is highlighting the underlying flaws in the way cities have been planned and run.

A. Cheshmehzangi (✉) · A. Sharifi
School of Architecture, Design and Planning (ADP), The University of Queensland, Brisbane, Australia
e-mail: A.Chesh@uq.edu.au

J. Zuo
School of Civil Engineering and Construction Management, Adelaide University, Adelaide, Australia

A. Sharifi
The IDEC Institute and Network for Education and Research on Peace and Sustainability (NERPS), Hiroshima University, Hiroshima, Japan

A. Cheshmehzangi · R. Zhang
School of Architecture and Planning, Hunan University, Changsha, China

A. Ziafati Bafarasat
School of the Built Environment, Oxford Brookes University, Oxford, UK

J. Zhao
The Weitzman School of Design, The University of Pennsylvania, Philadelphia, USA

P3Global, Dubai, UAE

A. Cheshmehzangi et al. (eds.), *Nature-based and Climate Responsive Design for Urban Systems*, Urban Sustainability, https://doi.org/10.1007/978-981-92-2070-0_1

This book's central thesis is that urban systems cannot continue to operate as closed, artificial entities that are isolated from ecological processes. Rather, they need to be rethought as flexible, open systems that function both inside and outside of natural dynamics. This is a practical necessity rather than just a philosophical change. Coastal inundation, protracted heatwaves, intensifying rainfall events, and environmental degradation are now defining characteristics of urban life rather than incidental threats [9]. In the face of compound and cascading climate consequences, conventional responses, which are usually focused on increasing gray infrastructure, are proven to be both functionally and economically unsustainable [8].

Infrastructure is not the only issue while the reasoning behind it is also problematic [3, 7]. Conventional urban systems are built with predictability and linear performance in mind: a transport network is optimized for anticipated flows, a seawall is built for a defined storm surge, and a drainage system is calculated for a certain rainfall event. However, by bringing in non-linearity, unpredictability, and extremes, climate change challenges these presumptions. Stability-oriented systems are increasingly faced with variability that they were never meant to handle. Failure consequently becomes systematic rather than an anomaly.

Rather than pursuing an incremental improvement, this book presents nature-based and climate-responsive design as a reconfiguration of this underlying logic. The focus moves from single-function infrastructure to multi-functional systems, from static solutions to adaptive processes, and from opposing environmental influences to collaborating with them. In this sense, nature is essential to how urban systems function, adapt, and survive; it is neither an aesthetic element nor a secondary layer of urban planning.

The understanding of water as a structuring agent in urban systems is a crucial aspect of this change. Rainfall, surface runoff, groundwater, and coastal dynamics are all examples of how water exposes the shortcomings of traditional design. For example, judgments that have prioritized impermeability, speed of discharge, and spatial efficiency over absorption, retention, and ecological balance have contributed to flooding in densely populated areas. In a similar vein, the elimination of flora and the predominance of hard surfaces that disturb local microclimates exacerbate heat stress. These are systemic consequences of the planning and construction of cities, not isolated technical problems.

Therefore, the subject of how urban systems might be constructed to absorb, adapt, and co-evolve with environmental conditions rather than collapse under them is directly addressed in the book. The response is examined through a series of case-based studies that show how nature-based tactics can be integrated at various scales, from more expansive landscape and coastal systems that redefine urban resilience to ground-level interventions like vegetative cover and decentralized drainage [1]. These are situational reactions influenced by socioeconomic, climatic, and cultural circumstances rather than idealized instances.

Crucially, the debate transcends the "gray versus green" infrastructure dichotomy. The difficulty lies in creating hybrid systems where ecological processes improve, supplement, and sometimes even redefine infrastructure performance rather than substituting one for the other. This necessitates a different approach to design, one that puts flexibility, redundancy, modularity, and interdependence ahead of control and optimization. Recognizing that urban systems are fundamentally unequal due to differences in resources, governance, and risk exposure is also necessary. Many of the most creative and essential solutions are coming from areas and places where limitations call for resource-efficient and adaptable solutions rather than from environments with abundant resources.

In this sense, the climate age is a time of both crisis and recalibration. It compels a re-evaluation of urban resilience's definition, methods, and target audience. By presenting nature-based and climate-responsive design as practical approaches for rethinking urban systems that are founded on practical applications and influenced by various geographical locations, this book helps with that recalibration. Instead of offering a universal paradigm, the goal is to outline a path toward cities that learn to coexist with nature rather than trying to outsmart it.

1.1 From Control to Coexistence: Positioning Nature-Based Design in Urban Systems

The developing alternative is a fundamental repositioning toward coexistence rather than just greater control, if the failure of traditional urban institutions is due to their reliance on control. In this sense, nature-based design is a reorganization of how urban settings are planned, structured, and run rather than an additional layer added to current systems. It suggests that resilience is attained by alignment with ecological processes, challenging the notion that resilience can be manufactured through resistance alone.

The understanding that natural systems are active, dynamic actors that influence spatial, environmental, and social consequences rather than passive backdrops to urban development is at the heart of this change. Soils control water cycles, rivers flood, coastlines shift, and vegetation controls temperature. These processes are functions that need to be comprehended, accommodated, and integrated rather than disruptions that need to be removed. Therefore, nature-based design reframes urban issues like heat stress, flooding, and biodiversity loss as signs of a breakdown between built form and biological processes rather than as discrete technological issues.

When water and landscapes are treated as infrastructure systems, this repositioning is more noticeable. Nature-based techniques prioritize retention, infiltration, and delayed release over rushing water away from cities through pipelines and channels, enabling urban environments to absorb and regulate unpredictability. In this way, blue-green infrastructure functions as a physical and organizational framework that transforms the way urban places operate in addition to being an environmental intervention. These days, streets, parks, and public areas serve multiple purposes at once, supporting ecological health, recreation, mobility, and water management.

The transition from defensive to adaptive design logics is equally significant. Conventional infrastructure is usually built with a predetermined threshold in mind;

failure happens when this barrier is exceeded. Nature-based systems, on the other hand, are naturally scalable and adaptive, able to respond to a variety of circumstances without failing. For example, a wetland responds to shifting water levels by expanding, contracting, and adjusting rather than "failing" like a drainage pipe does. In a climate situation marked by uncertainty and extremes, this ability to respond gradually rather than in a binary fashion is crucial.

However, placing wildlife in urban systems is a social and cultural issue in addition to a technical or ecological one. Communities' interactions with, appreciation of, and reliance on their ecosystems must be considered in nature-based design. Relationships with land and water are profoundly ingrained in everyday life and cultural practice in many contexts, especially in the Global South and tiny island regions. Therefore, incorporating nature into urban systems requires acknowledging local knowledge, customs, and goals in addition to physical design ideas. This supports the notion that effective remedies must be context-specific and socially anchored rather than being globally applicable.

Here, the idea of nature as infrastructure takes center stage. It goes beyond the concept of ecosystem services to present ecological systems as essential elements of urban functionality, on par with networks of water, electricity, and transportation [2, 6]. This necessitates reconsidering planning and governance frameworks so that nature is viewed as vital infrastructure that requires investment, upkeep, and long-term integration rather than as residual space or amenity.

Ultimately, the shift from control to coexistence is an active design strategy rather than a passive acceptance of environmental influences. Indeed, intentionality plays a crucial role in space allocation, system configuration, and performance assessment. Cities may develop resilience that is not just stronger but also more equal, flexible, and sensitive to the challenges of the climate age by incorporating ecological logic into urban systems.

2 Structuring the Argument: Two Complementary Pathways of Nature-Based Urbanism

This book is purposefully divided into two parts. It shows a conceptual evolution from embedded, site-level interventions to systemic, inter-scalar transformations in addition to thematic grouping. When combined, these two approaches present nature-based urbanism as both strategically revolutionary and practically feasible, answering the need for both short-term action and long-term urban system reconstruction.

"Part I: Water-Sensitive and Community-Scale Nature-Based Urban Systems" focuses on the neighborhood, street, building interface, and public realm as the scales where urban change is most noticeable. The justification for starting at this point is simple. Communities are most immediately affected by the effects of climate stress at this scale, including flooding, heat, water scarcity, and environmental degradation.

Additionally, it's where interventions may be gradually put into practice, tested in real-world settings, and modified over time.

The chapters in this part highlight water as a key organizing factor. Urban shape and function can be redefined by this resource and structuring force, not as an issue that needs to be eliminated. Decentralized drainage, water-sensitive urban design, vegetative ground cover systems, and blue-green public spaces are examples of approaches that are investigated as integrated methods that link social and spatial outcomes with environmental performance.

The community-oriented aspect is equally significant. These chapters demonstrate how nature-based solutions are integrated into daily urban life and are not just technical implementations. Accessible, inclusive, and locally flexible solutions are needed in public areas, informal settlements, and quickly urbanizing situations. The emphasis on emerging and climate-vulnerable areas is especially important since it casts doubt on the notion that high-resource places are the primary sources of innovation. Rather, it shows that resource limitations frequently encourage more resilient and context-sensitive design solutions.

In contrast, "Part II: Ecological Systems, Coastal Resilience, and Inter-Scalar Urban Futures" expands the conversation to include more significant territorial and systemic issues rather than just the immediate effects of site-based actions. This is justified by the fact that many of the most urgent climate issues, especially those pertaining to coastal exposure, ecosystem degradation, and regional environmental change, cannot be solved at the local level alone. From isolated locations to urban areas and biological landscapes, they necessitate cross-scale collaboration.

This part introduces the concept of inter-scalar design, where urban systems are understood as nested within broader environmental systems. For instance, coastal towns' resilience depends on integrated systems that incorporate mangroves, wetlands, coastal geomorphology, and urban form rather than just isolated interventions or built barriers. In a similar vein, the concept of "rewilding urban regions" goes beyond greening projects to take into account the possibility of reestablishing and connecting ecological networks throughout dispersed urban areas.

The framing of nature as infrastructure is a major focus in this section. This causes the conversation to change from considering ecological components as auxiliary or supplemental to acknowledging them as the main systems that promote urban resilience. Because ecological infrastructure requires different types of management than conventional physical systems, it also poses important problems regarding long-term planning, investment, and governance.

The two components are not independent even though they deal with different issues and function at different scales. Large-scale ecological methods must be implemented locally to be effective, just as site-based interventions become important when they support larger systemic change. As a result, the book's structure represents a continuum, ranging from instant application to long-term transformation, and from embedded practices to systemic frameworks. The advancement of nature-based urbanism as a design methodology and a strategy paradigm for climate-responsive urban systems depends on this dual viewpoint.

3 Aims, Objectives, and Contributions: Bridging Research, Policy, and Practice

The goal of this book is to move nature-based and climate-responsive design from a supporting role to a fundamental framework for urban systems. By providing a grounded but forward-looking contribution that is both analytically sound and practically applicable, it seeks to close enduring gaps between research, policy, and practice.

Fundamentally, the book aims to accomplish three interconnected goals. In the first place, it fosters a better comprehension of nature-based design as a systemic approach as opposed to a collection of discrete interventions. The book shows how these tactics might alter urban performance at various scales by placing case studies within larger environmental, social, and spatial dynamics. Second, through a variety of case studies that represent a range of meteorological conditions, socioeconomic realities, and governance contexts, it offers empirically grounded ideas. These examples are offered as situational solutions that provide transferable concepts and flexible frameworks rather than as universal models. Third, it seeks to convert conceptual discourse into practical knowledge so that those engaged in design, implementation, and decision-making can access the insights.

The book is highly relevant to policy development, especially in places where cities are struggling with resilience planning and climate adaptation. It adds to the continuing conversations on how land-use planning systems, coastal protection frameworks, and urban water management methods might incorporate nature-based solutions. Crucially, it emphasizes the necessity of policy approaches that transcend sectoral silos, acknowledging that integrated, cross-disciplinary collaboration is necessary for effective climate solutions. The book offers a foundation for better-informed policy frameworks that are in line with larger sustainability goals and sensitive to local realities by utilizing case-based data.

The book provides specific design and planning ideas that can guide practical implementations from a practice-oriented standpoint. These include methods for coordinating ecological processes with urban form, decentralized water management systems, and blue-green infrastructure integration into dense urban areas. The focus is on adaptation and feasibility, considering the limitations that many urban projects face. Instead of offering rigid solutions, the book offers adaptable strategies that may be adjusted to various scales and situations.

This work's integrated and multi-scalar perspective is what makes it novel. It illustrates the connections between regional policies and local actions by combining larger ecological systems and community-scale interventions into a single framework. Additionally, it highlights situations that are frequently overlooked in popular discourse, especially cities and areas with low resources and severe climate stresses. By portraying nature-based and climate-responsive design as crucial to the future of urban systems rather than optional additions, the book goes beyond descriptive descriptions to provide a directional contribution.

4 From Framing to Application: Setting the Stage for Case-Based Inquiry

One inevitable conclusion emerges from the ideas presented in this chapter: urban systems need to be redefined in relation to natural processes rather than in opposition to them [4]. The limitations of traditional planning and design methods that depend on control, forecasting, and infrastructure domination have been made clear by the climatic period. Instead of a single solution, a reorientation of practice that acknowledges nature as a system to engage rather than suppress and uncertainty as a condition to work with emerges.

Nature-based and climate-responsive design has been presented in this book as an operational approach as well as a conceptual change. It has made clear that incorporating ecological systems into urban settings is a required reaction to problems that are becoming more complicated and interconnected rather than an idealistic goal. However, the practical application of this repositioning is just as valuable as its theoretical consistency. The true test of these concepts is how they are applied, modified, and translated in various circumstances.

Therefore, the subsequent chapters are essential to the book's contribution and are not merely illustrative add-ons. Each case study addresses particular issues, such as ecological restoration, public space design, water management, or coastal vulnerability, and shows how nature-based methods may be implemented, negotiated, and integrated into actual urban systems. Crucially, these examples do not offer idealized or broadly applicable models. Rather, they are a reflection of situated behaviors that are influenced by environmental realities, cultural dynamics, local restrictions, and governance frameworks. In order to shift the conversation from abstraction to application, this grounded viewpoint is crucial.

The interaction of scale, context, and adaptability is a major theme that permeates the following chapters. The examples in Part I demonstrate how carefully planned, comparatively small-scale interventions can have cumulative effects that go beyond their immediate impact. They show how community-focused and water-sensitive strategies may transform ordinary urban settings, making them more adaptable, inclusive, and resilient. The scenarios in Part II, on the other hand, broaden the lens to broader systems, where tackling coastal threats, landscape fragmentation, and long-term environmental change requires the integration of ecological infrastructure and inter-scalar thinking. When taken as a whole, they support the notion that urban resilience is attained through links across scales rather than at a single scale.

The significance of design as a mediating technique is another crucial realization that supports the case studies. Nature-based solutions necessitate careful balancing of human demands, spatial configurations, and environmental processes; they are not merely technical additions. This establishes design as a link between engineering, ecology, planning, and social participation. It also emphasizes how important it is for the design process to be flexible, iterative, and learning-oriented, especially in situations where uncertainty is high and conditions are always changing.

As the reader progresses through the subsequent chapters, the focus switches from defining the issue to examining practical solutions. The deliberate diversity of cases provides several points of entry into the larger conversation on nature-based urbanism. When taken as a whole, they provide a multi-layered understanding of how cities might start to shift from systems of resistance to systems of coexistence and adaptability. The foundation has been laid in this chapter. What follows is the evidence.

References

1. Alhasaani ZS, Abdulameer L, Nile BK (2025) Urban land use change and sewer system resilience: a comprehensive review. Innov Infrastruct Solut 10(12):545
2. Gómez-Baggethun E, Gren Å, Barton DN, Langemeyer J, McPhearson T, O'farrell P, Andersson E, Hamstead Z, Kremer P (2013) Urban ecosystem services. In: Urbanization, biodiversity and ecosystem services: challenges and opportunities: a global assessment. Springer Netherlands, Dordrecht, pp 175–251
3. Herman R, Ausubel JH (eds) (1989) Cities and their vital systems: infrastructure past, present, and future. National Academies Press, Washington D.C.
4. Hough M (2002) Cities and natural process. Routledge, New York and London
5. IPCC (2022) Climate change 2022: impacts, adaptation and vulnerability. Cambridge: Cambridge University Press. Available online from: https://www.ipcc.ch/report/ar6/wg2/
6. Jansson Å (2013) Reaching for a sustainable, resilient urban future using the lens of ecosystem services. Ecol Econ 86:285–291
7. Pandit A, Minné EA, Li F, Brown H, Jeong H, James JAC, Newell JP, Weissburg M, Chang ME, Xu M, Yang P (2017) Infrastructure ecology: an evolving paradigm for sustainable urban development. J Clean Prod 163:S19–S27
8. UN-Habitat. (2020). World cities report 2020: the value of sustainable urbanization. United Nations Human Settlements Programme. Available online from: https://unhabitat.org/world-cities-report-2020-the-value-of-sustainable-urbanization
9. Young AF (2015) Extreme events and their impacts on urban areas. In: Young AF (ed) The Routledge handbook of urbanization and global environmental change. Routledge, New York and London, pp 229–246

Part I
Water-Sensitive and Community-Scale Nature-Based Urban Systems

Designing Nature Based Solutions for Urban Village Regeneration

Ying Zheng and Greg Keeffe

1 Introduction

Urban villages (Chengzhongcun in Chinese), often defined as informal urban settlements, have emerged as a distinctive and widespread spatial typology in the context of China's rapid urbanization and institutional transformation [13]. Characterized by extreme population density, compact building arrangements, low green coverage, air pollution, and inadequate infrastructure, these enclaves represent both a socio-spatial phenomenon and an environmental vulnerability [29, 30, 37, 38]. From a developmental perspective, urban villages have made significant positive contributions during China's rapid urbanization processes, particularly by accommodating large influxes of rural migrants with affordable housing and access to economic opportunities [32]. In cities such as Shenzhen, these informal settlements continue to function as critical entry points for migrant workers, providing low-cost living environments within otherwise unaffordable urban cores [18]. However, the very characteristics that make high-density urban villages accessible and affordable, limited public space, and substandard infrastructure, also generate serious environmental challenges (Fig. 1).

One of the most pressing environmental issues facing urban villages is the Urban Heat Island (UHI) effect [37]. The dense built form, limited vegetation, narrow alleys, and extensive use of impermeable surfaces result in increased ambient temperatures, reduced thermal comfort, and greater vulnerability to heat-related health problems [38]. These conditions are exacerbated during the summer months in subtropical regions such as the Pearl River Delta [35]. While cities around the world are turning to Nature-Based Solutions (NbS) to mitigate UHI and enhance climate resilience, the potential of NbS in the high-density informal settlement remains underexplored—both in the research and in practice.

Y. Zheng (✉) · G. Keeffe
Department of School of Natural and Built Environment, Queen's University Belfast, Belfast, UK
e-mail: y.zheng@qub.ac.uk

A. Cheshmehzangi et al. (eds.), *Nature-based and Climate Responsive Design for Urban Systems*, Urban Sustainability, https://doi.org/10.1007/978-981-92-2070-0_2

Fig. 1 An urban village in Shenzhen City. *Source* Baidu Map

This chapter addresses the urgent need for context-sensitive and design-driven Nature-based Solutions (NbS) in the regeneration of urban villages, with a particular emphasis on mitigating the Urban Heat Island (UHI) effect and enhancing living environment in the urban villages. It argues that the concept of Urban Forests as Nature-based Solutions (UF-NbS) provides a promising approach to achieving more sustainable urban living environments in the high-density urban areas. By integrating ecological thinking with spatial innovation, a transformative NbS strategy can be developed that simultaneously improves environmental liveability.

2 Nature-Based Solutions Strategy

Nature-based Solutions (NbS) can address the challenges of UHI and heat waves through the provision of cooling services [28]. By harnessing natural processes and ecosystem services, NbS contribute to environmental resilience and sustainability of human settlements while simultaneously improving the ecological integrity of surrounding landscapes [14]. Urban villages in China face complex and interrelated socioeconomic and environmental pressures, positioning them as key areas for transformative change. In this context, designing NbS can significantly enhance cities' adaptive capacity—not only by improving environmental resilience, but also by fostering social cohesion, place-making, public health (both physical and mental), and economic stability.

NbS strategies in high-density built environments such as urban villages must be carefully adapted to limited space, fragmented land ownership, and mixed-use functions. The typology of NbS in these contexts can be broadly categorized into

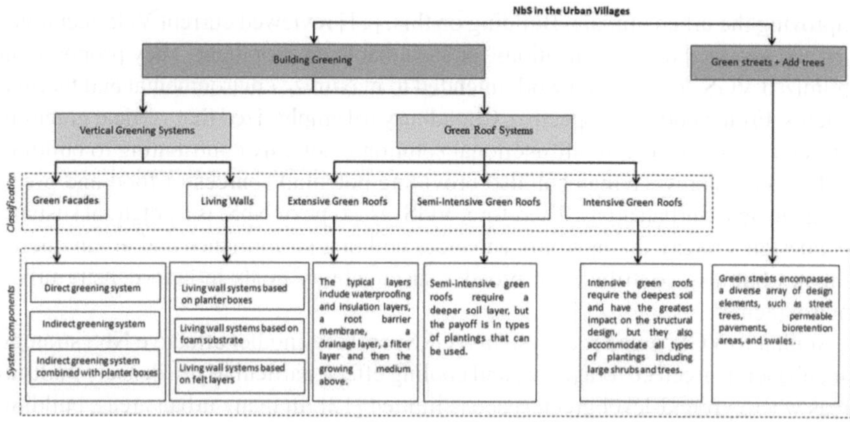

Fig. 2 Outline of the key elements of the conceptual framework for nature-based solutions. *Source* author

three systems: building greening (including vertical greening systems and green roofs systems) and green street interventions (Fig. 2).

2.1 Building Greening

Building greening stands out as an important example of NbS that can be effectively applied in densely urban areas, where green spaces are limited, and the built environment dominates. Building greening encompasses both vertical greening systems (VGS) and green roof systems (GRS), which can be flexibly integrated into existing building envelopes [21]. Vertical greening systems (VGS) are an integral component of sustainable building design, combining climbing plants, support structures, or modular planter boxes positioned at various heights to provide shading and thermal regulation [15, 22]. More advanced solutions, such as living wall systems, incorporate pre-vegetated panels that allow for a wider range of plant species. These systems contribute to passive cooling, improve building envelope performance, and reduce the urban heat island (UHI) effect. VGS have been shown to lower façade surface temperatures during summer and enhance insulation performance in winter [4]. By modifying the microclimate within dense urban areas—particularly in street canyons—VGS can reduce ambient temperatures through plant evapotranspiration processes. Numerous studies have investigated the thermal and environmental performance of VGS in urban contexts. For instance, Wong [34] evaluated eight different VGS types installed in Singapore's HortPark, examining their effects on both surface and ambient temperatures to assess thermal performance at the building and microclimate scales. Similarly, Perini et al. [23] highlighted the significant role of both vertical and horizontal greening in enhancing the thermal behaviour of buildings during both summer and winter seasons, as well as their broader contribution to improving the urban climate. Building on this, [11] reviewed current VGS technologies and assessed their implications for sustainable development. They proposed an optimized VGS design framework intended to maximize environmental and thermal benefits. From another perspective, Choudhury [6] emphasized that vertical greening can serve as a low-cost, multifunctional solution—not only contributing to building cooling and energy savings but also providing potential sources of food and green energy in dense urban areas. Therefore, VGS, as a type of NbS, is a potential sustainable design strategy that not only enhances building performance and urban microclimates but also contributes to broader sustainability goals in high-density urban environments.

Similar to VGS, green roof systems (GRS) provide another effective NbS strategy for enhancing green coverage rate and cooling effect, particularly in densely built-up areas where ground-level green space is limited [12]. In many urban areas, building roofs account for around 40–50% of the total impervious surface area [27]. These roofs have substantial potential for retrofitting buildings with roof greening for urban sustainable development [2]. Green roof systems (GRS) refer to roofing systems that incorporate vegetation, growing media (soil or substrate), and drainage layers on the top of buildings [3, 39]. Green roofs are generally divided into extensive and intensive systems, while some classifications also include semi-intensive types as an intermediate category [26, 37]. As a key component of NbS strategies to mitigate the UHI effect, GRS provide significant potential for improving microclimatic conditions in densely built environments. Several studies have examined the effectiveness of green roof systems (GRS) in mitigating the urban heat island (UHI) effect across different climatic contexts. In Southern California, Roche et al. [17] tested a hybrid green roof that integrates evaporative and radiant cooling systems, demonstrating its enhanced cooling performance in hot, arid climates and its applicability in projects across the US, Middle East, and Asia. Besides, Wang et al. [31] used the WRF/UCM and RayMan models to compare cool and green roofs, finding that highly reflective cool roofs outperformed irrigated green roofs in night-time cooling and thermal comfort improvement in Germany. In Beijing, Feng et al. [8] developed ENVI-met simulations as a method to assess green roofs in residential areas with varied building heights. The results showed modest pedestrian-level cooling effects, particularly on the downwind side, though their performance declined under extreme heat conditions. Thus, these findings suggest that the differences in green roof system design can cool the building under different climatic conditions and roof configurations.

2.2 Green Streets

Green streets, as a form of NbS, are typically implemented within the public right-of-way to intercept and treat polluted stormwater runoff from impervious surfaces before it is discharged into local waterways [10]. Green streets encompass a diverse range of design elements, such as street trees, permeable pavements, bioretention areas, and swales. Street trees and grass are important elements for cooling cities. In all design elements, adding trees and plants helps in the reduction of heat by providing shade and decreasing the amount of direct sunlight reaching the ground [16]. Coutts et al. [7] found that street trees can reduce summer daytime air temperatures by up to 1.5 °C; however, their cooling effect during extreme heat events showed limited differences between streets with and without tree cover. Ananyeva and Emmanuel [1] found that the "Avenues Programme"—a plan to convert all city centre thoroughfares into tree-lined streets—could significantly enhance thermal comfort during heatwaves, shifting conditions from the "hot" to the "slightly warm" category across the city centre. Building on this, Wu et al. [36] evaluated 32 design scenarios and identified optimal configurations of street trees and grass to maximize cooling effects in European neighbourhoods with temperate climates. These findings highlight the potential of green streets to serve as multifunctional urban cooling strategies.

3 Case Study

Baishizhou is one of Shenzhen's most densely populated urban villages, comprising five interconnected villages (Fig. 3) [33]. These include three former urban villages—Baishizhou, Shangbaishi, and Xabaishi—and two administrative villages, Xintang and Tangtou [37]. Four of these are located north of Shennan Road, a major east–west artery in the city. The boundaries between these informal settlements are loosely defined, often marked only by streets or address signs, and spatial differences among them are minimal. Baishizhou's urban morphology is characterized by three distinct architectural typologies: handshake buildings, factory compounds, and traditional tile-roofed houses [37]. This chapter explores how multi-scales, link-based design strategies can be deployed to address the challenges of UHI and support sustainable regeneration. The central research question guiding this investigation is: "How can Nature-based Solutions be designed to mitigate the Urban Heat Island (UHI) effect and improve thermal comfort in the urban villages?"

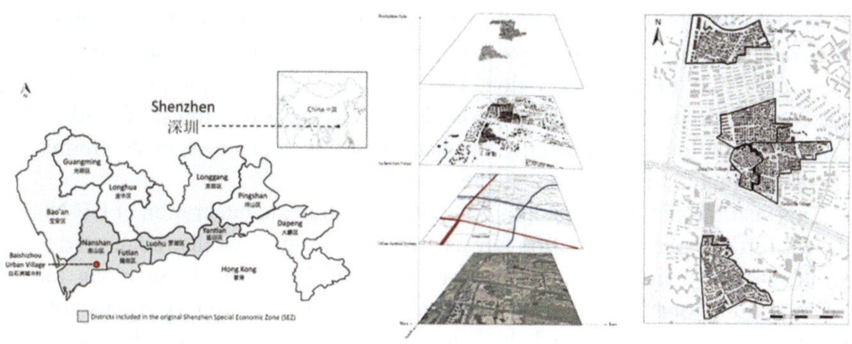

Fig. 3 Distribution map of the Baishizhou urban village. *Source* by authors, based on Baidu Map

4 Methodology: Research by Design

This study adopts research-by-design approach, which positions design as both a process and a method of inquiry. Rather than separating theory from practice, this methodology integrates the act of designing with critical reflection to generate new knowledge [24]. Design tools such as sketches, spatial mappings, 3D modelling, and iterations serve as research material, enabling a continuous dialogue between conceptual exploration and material testing. Within the field of architecture, research-by-design aligns creative practice with academic investigation, emphasizing that the design process itself can reveal new urban perspectives and challenge conventional development paths [5, 9]. It involves phases such as site analysis, problem framing, programme development, and iterative prototyping, facilitating both analytical and speculative thinking. This approach not only complements traditional research designs—such as case study or comparative analysis, but also enables the formulation of innovative, context-specific design responses [25]. Therefore, the study proposes the use of NbS as an alternative to demolition and reconstruction, with a focus on designing an "Urban Forest" to mitigate the UHI effect and improve thermal comfort.

5 Design Framework

Urban forests as nature-based solutions (UF-NbS) refer to tree-based urban ecosystems that are strategically designed to address societal challenges while delivering vital ecosystem services for human well-being and biodiversity conservation [20]. These solutions encompass a broad spectrum of interventions—from street trees and vertical forests to green corridors and urban parks—and are closely linked to ongoing efforts to expand, restore, and maintain urban green infrastructure. As such, UF-NbS are positioned as critical responses to pressing urban issues including climate change, air pollution, and ecological degradation.

In high-density urban contexts such as urban villages, the built environment is often described as a "Concrete Forest"—an agglomeration of hard surfaces, limited open space, and minimal vegetation. This raises an urgent design question: What if this concrete forest could be transformed into a living urban forest? Could such a transformation improve liveability, mitigate environmental stressors, and reimagine the spatial quality of dense settlements?

This vision is not merely utopian. Through the strategic integration of UF-NbS—such as vertical greening, rooftop forests, tree-lined streets, and interconnected forest neighbourhoods—it is both feasible and scalable to retrofit dense urban fabrics into multifunctional, climate-responsive environments. Such interventions not only address ecological and thermal challenges but also enhance social cohesion, public health, and spatial equity. Therefore, the transformation from "Concrete Forest" to

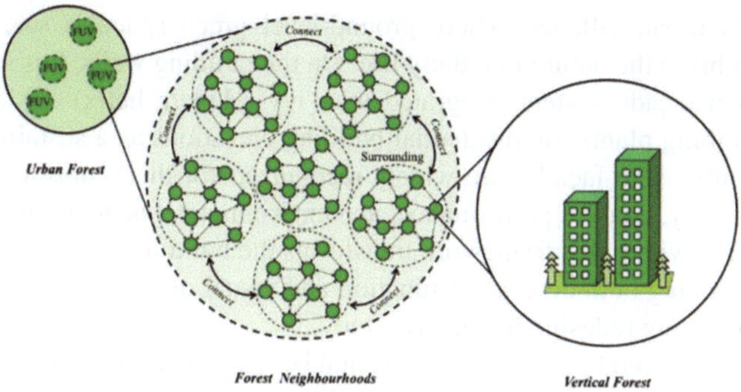

Fig. 4 Urban biosphere. This is a diagram of Urban Forest–Forest Neighborhood–Vertical Forest. *Source* by authors

"Urban Forest" can represent a realistic and actionable design strategy for future urban resilience.

The urban forest concept is structured across three interconnected spatial dimensions, each corresponding to a distinct scale of nature-based intervention. Specifically, three NbS strategies are proposed: vertical forests at the building scale, neighbourhood forest at the neighbourhood scale, and a broader urban forest at the district scale. These multi-layered strategies are designed to work in synergy to mitigate the UHI effect, enhance environmental quality, and improve living conditions in high-density urban areas. By embedding ecologically functional green networks within the existing urban fabric, the framework aims to restore environmental performance while maintaining spatial continuity and urban intensity (Fig. 4).

5.1 *Vertical Forest*

This study proposes an innovative "Vertical Forest" concept aimed at integrating natural elements into urban villages, where conventional tree planting is often unfeasible due to limited ground-level space. Instead, the strategy utilizes green façades and green roofs as scalable and sustainable alternatives to enhance ecological performance and mitigate the negative impacts of urbanization.

The design process focuses on developing context-specific strategies, with particular attention to reducing building surface temperatures and improving outdoor thermal comfort. Applied to residential buildings. The vertical forest approach demonstrates its potential as a form of urban micro-forestry, providing both climatic benefits and an added ecological layer within the dense built fabric.

The Vertical Forest (VF) concept includes Roof Greening Systems (GRS) Design and Vertical Greening Systems (VGS) Design. On the rooftop, this design strategy explores the potential of rooftop farming as a NbS strategy to enhance ecological performance and foster social engagement in high-density urban area. In

Baishizhou's urban villages, where ground-level green space is scarce, rooftop farming can bring the nature into daily life. On the building walls, this study adopts a double-skin façade system integrated with ivy (Hedera helix)—a fast-growing and self-climbing plant—on residential building elevations as a sustainable design strategy. Double-skin façades are widely recognized for their capacity to improve energy efficiency, thermal performance, and environmental adaptability. The incorporation of ivy within this framework transforms the building envelope into a living surface, enhancing both ecological function and visual quality. Figure 5 shows the residential building redesign in the case study.

Ivy provides several passive environmental benefits, including natural insulation, shading, and temperature regulation of the building surface. These characteristics contribute to reduced cooling loads and overall energy demand. Furthermore, the use of stainless-steel cable systems as climbing supports allows for a durable and visually integrated façade solution, blending natural growth with contemporary architectural expression. This design was selected not only for its climatic performance but also for its low maintenance requirements and the rapid growth rate of climbers like ivy, making it particularly suitable for retrofitting high-density residential environments with limited green space at ground level.

5.2 Neighbourhood Forest

At the mesoscale, this study introduces the concept of "Neighbourhood Forest" as a key component of the broader urban forest framework. These neighbourhood-scale interventions establish a network of green spaces and ecological corridors that link vertical forests at the building scale with larger urban forest systems, forming a continuous and scalable green infrastructure network. Unlike the conventional idea of "Green Neighbourhoods", the "Neighbourhood Forest" model is rooted in the vertical forest strategy, extending its ecological logic across communities to create localized forest ecosystems within high-density urban environments (Fig. 6). To create comfortable public spaces that encourage residents to spend more time outdoors, enhancing the thermal environment in urban streets is critical. The high-density urban pattern prevalent in the case study, which results in reduced green spaces and vegetation as well as limited air circulation, is a leading factor in the deterioration of the outdoor thermal environment. The study posits that the "Neighbourhood Forest" concept represents the most effective solution for enhancing outdoor thermal comfort in urban villages. The "Neighbourhood Forest" concept is rooted in the urban forest concept, but it emphasizes the significance of creating and maintaining forests in urban areas at the local or neighbourhood level. The objective of the neighbourhood forest concept is to increase green space and biodiversity in densely populated areas

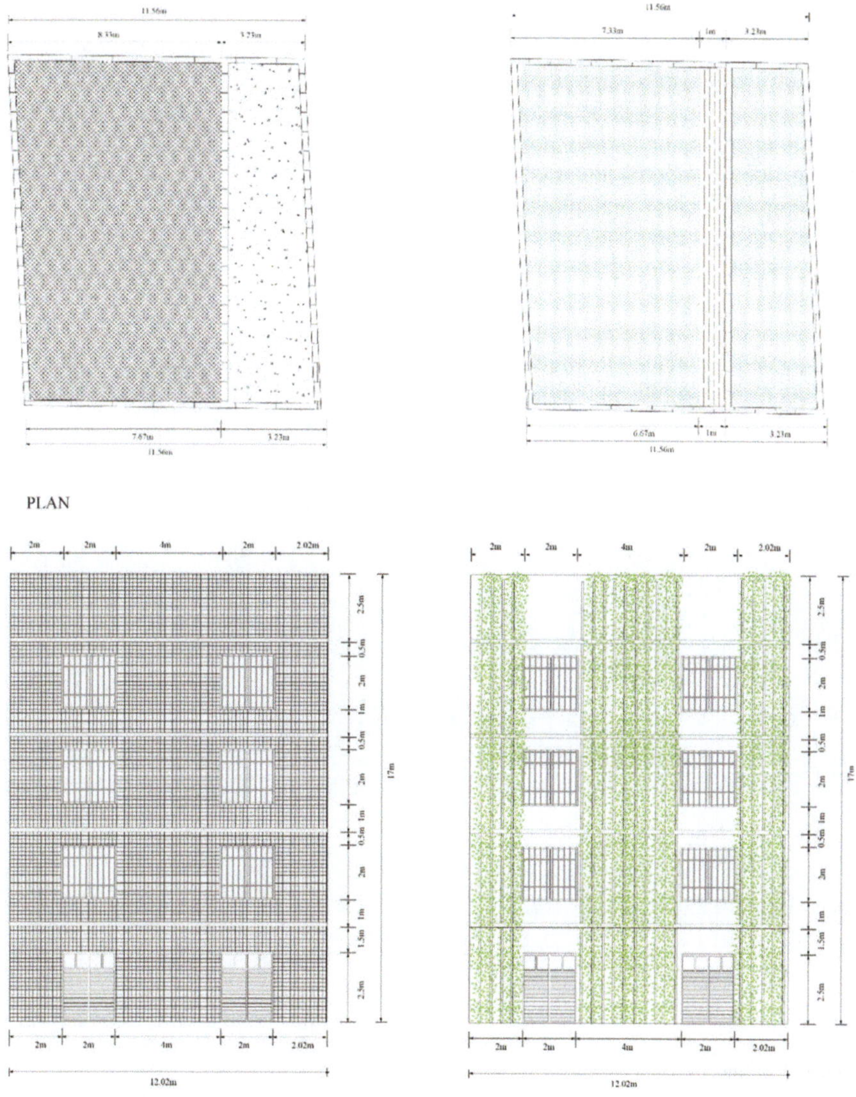

Fig. 5 Residential building redesign. *Source* by authors

and provide social and environmental benefits to nearby residents. The neighbourhood forests are regarded as a means of developing a nested network of green spaces within an urban landscape. By creating interconnected neighbourhood forests, it is possible to enhance the ecological connectivity of the urban landscape, allowing for the movement of flora and fauna and promoting biodiversity. This type of nested

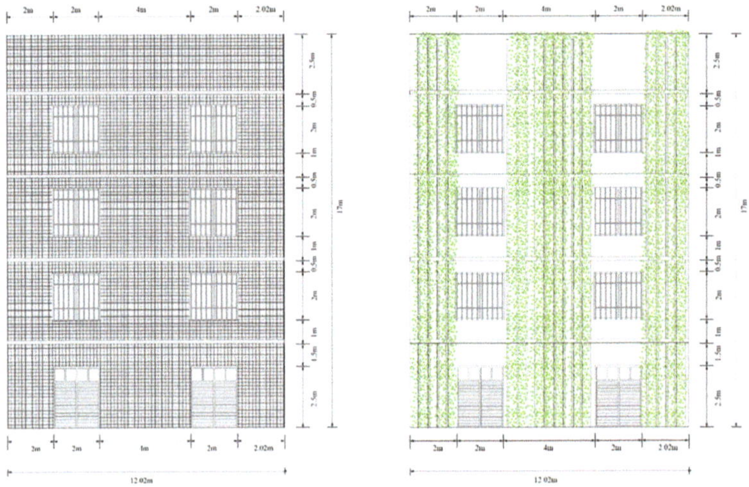

SOUTH ELEVATION

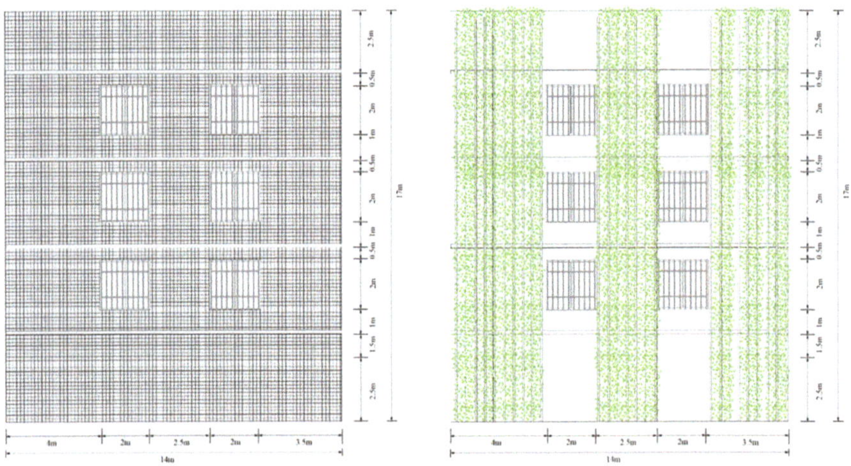

EAST ELEVATIOM

Fig. 5 (continued)

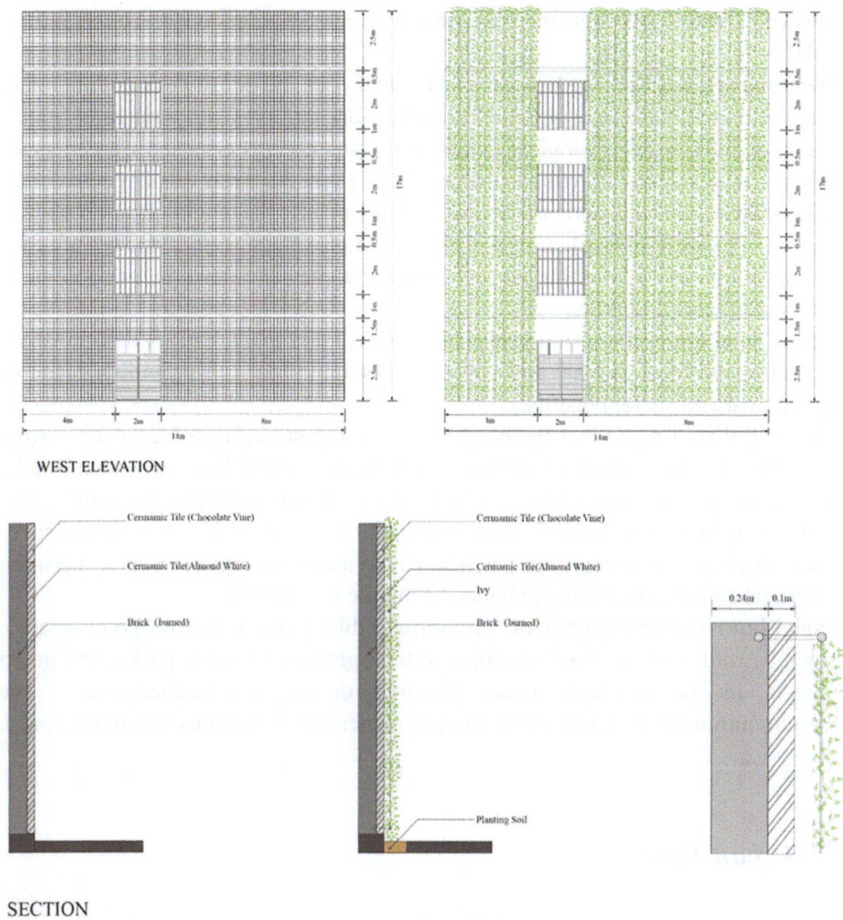

Fig. 5 (continued)

network emphasizes that landscape connectivity facilitates the movement of organisms, genetic interchange, and other ecological flows that are critical for the viability and survival of species and for the conservation of biodiversity in general [19]. Thus, this is in line with the principles of landscape ecology, which emphasizes the importance of ecological connections and interactions across different spatial scales. By developing neighbourhood forests as part of a broader network of green spaces within an urban landscape, it can be creating a more resilient and sustainable environment. Figure 6 presents a comparison between the original areas and the designated areas, focusing on the incorporation of neighbourhood forests. On the other hand, Fig. 7 illustrates the perspective of pedestrians, highlighting the contrast between the original areas and the designed areas. Building on the concept of neighbourhood forests, a "Neighbourhood Forest" design is proposed based on previous research. This design includes several key aspects:

- **Overall Green Coverage**: increasing areas' green coverage can effectively reduce UHI.
- **Neighbourhood Forest Structure**: plan the layout and density of trees and other vegetation to create an ecological forest network that provides a range of micro-habitats, maximizes ecological connectivity, and creates a sense of green space for nearby residents. It includes two aspects: (1) create green spots by retrofitting vacant land, idle land, and other available land areas at the neighbourhood scale to innovative green infrastructure designs; (2) improve connectivity of green space in urban blocks by stitching adjacent green patches and make them continuous green corridors and greenways.
- **Plant Selection**: choose a diverse range of tree and plant species that are well-suited to the local climate and soil conditions and that can provide a range of ecological and social benefits.
- **Accessibility**: Ensure that the neighbourhood forest is accessible in the narrow axle. Identify the crossing on the streets and create green axles between handshake buildings. It includes two aspects: (1) creates a continuous greenway, ensuring that the "Neighbourhood Forest" is accessible. At the same time, increases connectivity between adjacent green patches; (2) creates connective corridors between "vertical forest", increasing enhanced ecological corridors.
- **Neighbourhood Engagement**: providing public participation in green spaces, such as roof gardens, roof farming, vertical greening, pocket parks, and green streets. Involve residents in the planning, design, and maintenance of the neighbourhood forest, to build a sense of ownership and connection to the space.

5.3 Urban Forest

At the district scale, the proposed Urban Forest strategy synthesizes the preceding concepts of "Vertical Forest" and "Neighbourhood Forest", aiming to establish a cohesive green infrastructure network that enhances ecological connectivity and urban resilience. It acknowledges the significance of urban forests as a key component in addressing the multifaceted challenges of sustainable urban development and environmental management. Urban Forest (UF), as a subset of NbS, provides a viable approach to mitigate the adverse social-ecological consequences associated with urbanization, concurrently fostering human well-being and enhancing the urban environment. Figure 8 visually presents the NbS within the Baishizhou urban villages.

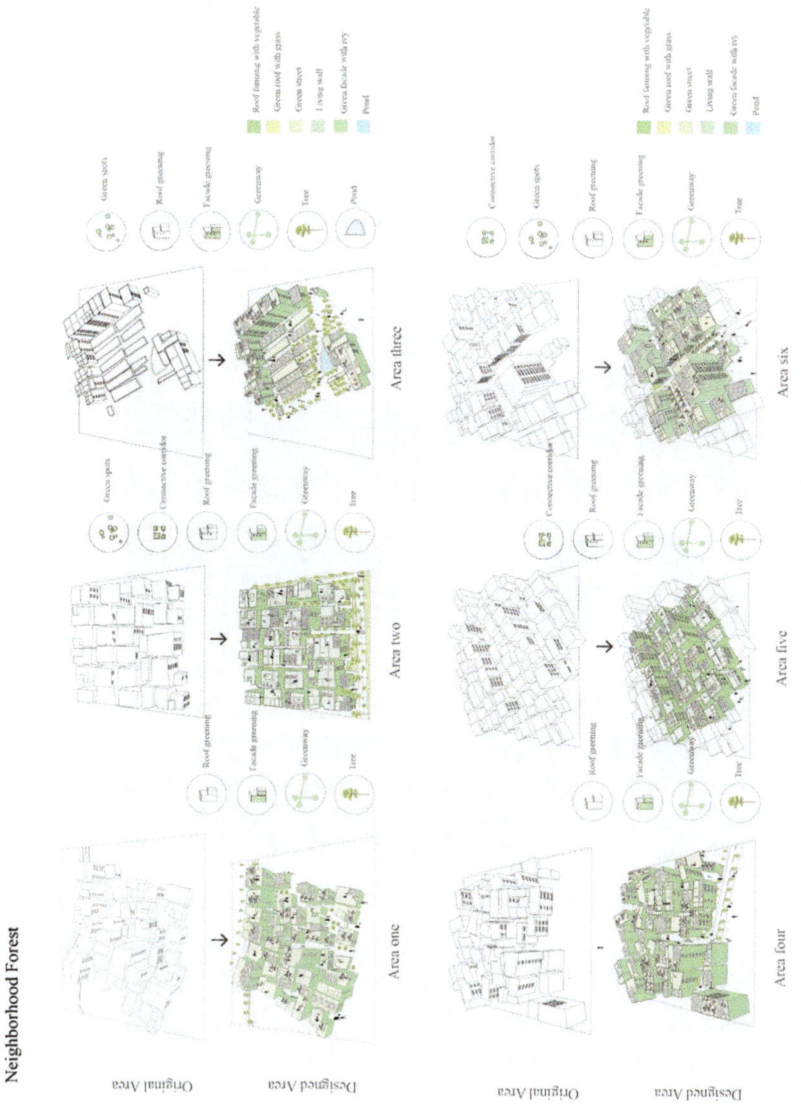

Fig. 6 The designed areas based on neighbourhood forest concept. *Source* by authors

Fig. 7 Pedestrian's perspective in the test area in different locations. *Source* by authors

Fig. 8 The design process for implementing urban forest-based solutions in diverse spatial contexts. *Source* authors

6 Discussion and Implications

This study takes Baishizhou—one of the largest and most controversial urban villages in Shenzhen—as a representative case of high-density informal settlements to explore the potential of "Urban Forest" as a "Nature-based Solution" (UF-NbS). Baishizhou is characterized by extremely compact building morphology, limited access to green spaces, and poor living conditions. Under the current urban regeneration policy, demolition-led redevelopment is the dominant strategy, aiming to replace the existing environment with high-rise commercial complexes. However, this top-down approach has raised serious concerns regarding environmental sustainability, social equity, and the loss of community identity.

In response, this research proposes an alternative regenerative vision grounded in the UF-NbS framework, which does not rely on destruction but on ecological integration and spatial layering. The proposed urban forest system forms a scalable ecological network that operates across three nested levels—"Vertical Forest" (building scale), "Neighbourhood Forest" (neighbourhood scale), and "Urban Forest" (district scale)—like a set of Russian dolls. These interlinked layers form a continuous urban ecological circle that enhances biodiversity, reduces the urban heat island (UHI) effect, and creates a healthier, more climate-resilient environment. At the building scale, green facades and rooftops improve indoor comfort and reduce energy demand by mitigating solar heat gain and enhancing thermal insulation. At the neighbourhood scale, tree-lined green streets and micro-forested public spaces enhance outdoor thermal comfort, particularly in street canyons and areas with limited air circulation. Finally, at the district scale, the cumulative effects of these interventions contribute to broader environmental benefits such as urban heat island (UHI) mitigation, increased biodiversity, and improved air quality.

Importantly, this design framework transcends the realm of utopian planning. It provides a practical and replicable solution grounded in ecological logic, adaptable to the spatial and environmental constraints of densely built environments. The modular and scalable nature of this NbS system supports its integration into real-world planning and architectural practices.

Moving forward, simulation-based testing and environmental performance analysis will be conducted to quantitatively evaluate the feasibility and impact of this approach, thereby bridging the gap between conceptual design and implementable urban policy.

7 Concluding Remarks

This chapter proposes an integrated Urban Forest design strategy as a nature-based solution for regenerating high-density informal settlements, using Baishizhou in Shenzhen as a representative case. Rather than adopting conventional demolition-led redevelopment, this research introduces a multi-scalar ecological framework—“Vertical Forest” (building scale), “Neighbourhood Forest” (neighbourhood scale), and “Urban Forest” (district scale)—to form a nested, forest-like urban ecosystem.

The design strategy aims to mitigate the UHI effect. Nature-based design is not an abstract utopia, but a practical, scalable strategy that can be implemented in dense urban contexts. Grounded in research-by-design methodology, this vision offers an alternative, sustainable path for urban transformation. Future work will focus on simulation and empirical evaluation to further test the feasibility and performance of the proposed solutions.

References

1. Ananyeva O, Emmanuel R (2023) Street trees and urban heat Island in Glasgow: mitigation through the “avenues programme”. Urban For Urban Green 86:128041. https://doi.org/10.1016/j.ufug.2023.128041
2. Berardi U, GhaffarianHoseini A, GhaffarianHoseini A (2014) State-of-the-art analysis of the environmental benefits of green roofs. Appl Energy 115:411–428. https://doi.org/10.1016/j.apenergy.2013.10.047
3. Besir AB, Cuce E (2018) Green roofs and facades: a comprehensive review. Renew Sust Energ Rev 82:915–939. https://doi.org/10.1016/j.rser.2017.09.106
4. Bustami RA, Belusko M, Ward J, Beecham S (2018) Vertical greenery systems: a systematic review of research trends. Build Environ 146:226–237. https://doi.org/10.1016/j.buildenv.2018.09.045
5. Ceschin F, Gaziulusoy I (2016) Evolution of design for sustainability: from product design to design for system innovations and transitions. Des Stud 47:118–163. https://doi.org/10.1016/j.destud.2016.09.002
6. Choudhury S (2021, October 15) Vertical greening is a low-cost source of food, energy and building cooling. JPI Urban Europe. https://jpi-urbaneurope.eu/vertical-greening-is-a-low-cost-source-of-food-energy-and-building-cooling/
7. Coutts AM, White EC, Tapper NJ, Beringer J, Livesley SJ (2016) Temperature and human thermal comfort effects of street trees across three contrasting street canyon environments. Theor Appl Climatol 124(1):55–68. https://doi.org/10.1007/s00704-015-1409-y
8. Feng Y, Liu L, Yang L, Wang Z (2022) Evaluating the cooling performance of green roofs under extreme heat conditions. Front Environ Sci 10. https://doi.org/10.3389/fenvs.2022.874614

9. Hauberg J (2011) Research by design—a research strategy. Arch & Educ J, January 2011, 46–56
10. Im J (2019) Green streets to serve urban sustainability: benefits and typology. Sustainability 11(22):6483. https://doi.org/10.3390/su11226483
11. Irga PJ, Pettit T, Torpy FR, Tan Z (2023) Vertical greening systems: a perspective on existing technologies and new design recommendation. Sustainability 15(7):6014. https://doi.org/10.3390/su15076014
12. Jamei E, Seyedmahmoudian M, Jarimi H, Thirunavukkarasu G, Stojcevski A (2021) Review on the cooling potential of green roofs in different climates. Sci Total Environ 791:148407. https://doi.org/10.1016/j.scitotenv.2021.148407
13. Kan K (2016) The transformation of the village collective in urbanising China: a historical institutional analysis. J Rural Stud 47:588–600. https://doi.org/10.1016/j.jrurstud.2016.07.016
14. Kauark-Fontes B, Marchetti L, Salbitano F (2023) Integration of nature-based solutions (NBS) in local policy and planning toward transformative change. Evidence from Barcelona, Lisbon, and Turin. Ecol Soc 28(2). https://doi.org/10.5751/es-14182-280225
15. Köhler M (2008) Green facades—a view back and some visions. Urban Ecosyst 11:423–436. https://doi.org/10.1007/s11252-008-0063-z
16. Kumar P, Abhijith KV, Barwise Y, Kershaw T, Morawska L (2024) Urban heat mitigation by green and blue infrastructure: drivers, effectiveness, and future needs. Int J Hydrogen Energ 5(2):100588. https://doi.org/10.1016/j.xinn.2024.100588
17. La Roche P, Yeom DJ, Ponce A (2020) Passive cooling with a hybrid green roof for extreme climates. Energ Buildings 224:110243. https://doi.org/10.1016/j.enbuild.2020.110243
18. Liu S, Zhang Y (2020) Cities without slums? China's land regime and dual-track urbanization. Cities 101:102652. https://doi.org/10.1016/j.cities.2020.102652
19. Mimet A, Houet T, Julliard R, Simon L (2013) Assessing functional connectivity: a landscape approach for handling multiple ecological requirements. Methods Ecol Evol 4(4):453–463. https://doi.org/10.1111/2041-210x.12024
20. Mosca F, Canepa M, Perini K (2023) Strategies for adaptation to and mitigation of climate change: key performance indicators to assess nature-based solutions performances. Urban Clim 49:101580. https://doi.org/10.1016/j.uclim.2023.101580
21. Perini K, Rosasco P (2016) Is greening the building envelope economically sustainable? An analysis to evaluate the advantages of economy of scope of vertical greening systems and green roofs. Urban For & Urban Green 20:328–337. https://doi.org/10.1016/j.ufug.2016.08.002
22. Perini K, Ottelé M, Haas EM, Raiteri R (2012) Vertical greening systems: a process tree for green façades and living walls. Urban Ecosyst 16:265–277. https://doi.org/10.1007/s11252-012-0262-3
23. Perini K, Ottelé M, Fraaij ALA, Haas EM, Raiteri R (2011) Vertical greening systems and the effect on air flow and temperature on the building envelope. Build Environ 46(11):2287–2294. https://doi.org/10.1016/j.buildenv.2011.05.009
24. Roggema R (2016) Research by design: proposition for a methodological approach. Urban Sci 1(1):2. https://doi.org/10.3390/urbansci1010002
25. Saunders MNK, Rojon C (2011) On the attributes of a critical literature review. Coaching 4(2):156–162. https://doi.org/10.1080/17521882.2011.596485
26. Shafique M, Kim R, Rafiq M (2018) Green roof benefits, opportunities and challenges—a review. Renew Sust Energ Rev 90:757–773. https://doi.org/10.1016/j.rser.2018.04.006
27. Stovin V (2009) The potential of green roofs to manage urban stormwater. Water Environ J 24(3):192–199. https://doi.org/10.1111/j.1747-6593.2009.00174.x
28. Su W, Zhang L, Chang Q (2022) Nature-based solutions for urban heat mitigation in historical and cultural block: the case of Beijing Old City. Build Environ 225:109600. https://doi.org/10.1016/j.buildenv.2022.109600
29. Tan X, Zhang Y, Lin Y, Zheng Y (2019) Localized place-making and the knowledge-based regeneration strategies: the case of Xiasha village in Shenzhen. Habitat Int 83:73–84. https://doi.org/10.1016/j.habitatint.2018.11.005

30. Van Oostrum M (2021) Access, density and mix of informal settlement: comparing urban villages in China and India. Cities 117:103334. https://doi.org/10.1016/j.cities.2021.103334
31. Wang X, Li H, Sodoudi S (2022) The effectiveness of cool and green roofs in mitigating urban heat Island and improving human thermal comfort. Build Environ 217:109082. https://doi.org/10.1016/j.buildenv.2022.109082
32. Wang YP, Wang Y, Wu J (2010) Private rental housing in 'urban villages' in Shenzhen: problems or solutions? In: Wu F, Webster C (eds) Marginalization in urban China: comparative perspectives. Palgrave Macmillan, pp 153–174
33. Wang Y, Kacmaz Erk G (2022) Rural-urban Chinese migration and an architectural investigation of the urban village in Shenzhen. The International Journal of Community Diversity 22(2):1–14. https://doi.org/10.18848/2327-0004/cgp/v22i02/1-14
34. Wong NH, Tan AYK, Tan PY, Wong NC (2010) Thermal evaluation of vertical greenery systems for building walls. Build Environ 45(3):663–672. https://doi.org/10.1016/j.buildenv.2009.08.005
35. Wu W, Zhang X, Zhu X, Chen Y (2018) Distinct influences of urban villages on urban heat islands: a case study in the Pearl River Delta, China. Int J Environ Res Public Health 15(8):1666. https://doi.org/10.3390/ijerph15081666
36. Wu Y, Mashhoodi B, Patuano A (2025) Effective street tree and grass designs to cool European neighbourhoods. Urban Clim 61:102376. https://doi.org/10.1016/j.uclim.2025.102376
37. Zheng Y (2023) Nature-based solutions for urban heat islands in megacities: a case study in Shenzhen, China (Doctoral thesis). Queen's University Belfast
38. Zheng Y, Keeffe G, Mariotti J (2023) Nature-based solutions for cooling in high-density neighbourhoods in Shenzhen: a case study of Baishizhou. Sustainability 15(6):5509. https://doi.org/10.3390/su15065509
39. Zheng X, Chen X, Guo W, Yu R, Zhang H (2021) Green roofs for stormwater runoff retention: a global quantitative synthesis of the performance. Resour Conserv Recycl 170:105577. https://doi.org/10.1016/j.resconrec.2021.105577

Inclusive Blue-Green Public Open Spaces: Learning from Tansen, Nepal

Ilija Gubic, Stefania Leontiadou, Anup Ghimire, and Manju Raut

1 Importance of Wetlands for Cities

Urbanization, one of the defining global trends of the twenty-first century, has profound implications for the natural environment and the quality of life for urban residents. With more than half of the world's population living in cities [47], urban planners and policymakers are increasingly focusing on integrating nature-based solutions into the urban fabric to promote sustainability, enhance ecological resilience, and improve public health outcomes in cities [25, 50, 52]. One such strategy is the conservation and rehabilitation of urban wetlands as part of green infrastructure systems [37, 52]. Urban wetlands, traditionally seen as wastelands or undeveloped land, are now being revalued for their multifaceted ecological, social, and health benefits [1, 30, 37, 52].

Green infrastructure refers to a network of natural and semi-natural spaces that provide ecosystem services and support sustainable urban living [18]. Wetlands, as an integral part of green infrastructure, offer a set of services, including flood attenuation, water purification, biodiversity support, and climate regulation, with studies showing that they contribute to stormwater management by acting as natural sponges, absorbing excess water during heavy rainfall and reducing the risk of flooding [31, 53]. Moreover, they play a key role in improving air and water quality by filtering pollutants, thereby enhancing the overall environmental quality of cities [31, 53].

I. Gubic (✉)
Faculty of Architecture, University of Belgrade, Belgrade, Serbia
e-mail: 2020_41012@edu.arh.bg.ac.rs

S. Leontiadou
School of Architecture, Aristotle University of Thessaloniki, Thessaloniki, Greece

A. Ghimire
Sunwal, Nepal

M. Raut
Tansen, Nepal

A. Cheshmehzangi et al. (eds.), *Nature-based and Climate Responsive Design for Urban Systems*, Urban Sustainability, https://doi.org/10.1007/978-981-92-2070-0_3

Research also highlights the significance of wetlands in carbon sequestration, making them vital components in efforts to combat global warming [53]. These benefits are especially critical in densely populated urban environments, where access to nature is often limited. By designing well-managed public open spaces around urban wetlands while safeguarding ecological functions, cities can provide residents with much-needed access to nature, contributing to more equitable urban environments [32].

The integration of wetlands into urban green infrastructure also aligns with broader sustainability goals. As cities grapple with the dual challenges of climate change and rapid urbanization, there is a growing recognition that sustainable urban development requires the preservation and enhancement of natural ecosystems [31]. Wetlands, with their ability to regulate local climates, contribute to microclimate moderation and urban resilience, and store carbon, are critical to achieving urban sustainability [18, 31, 53]. Additionally, they serve as important habitats for a wide range of species, contributing to urban biodiversity and supporting ecosystem resilience in the face of environmental stressors [17, 31].

The transformation of urban wetlands into wetland-integrated public open spaces, such as parks, reflects a paradigm shift in how cities approach environmental management and urban development. This shift acknowledges that wetlands provide critical ecosystem services, including flood regulation, water purification, and biodiversity support, while also offering recreational and mental health benefits to urban populations [23, 27, 50], helping to alleviate stress, anxiety, and depression [7, 22]. Such parks offer opportunities for outdoor recreation, including walking, birdwatching, and other leisure activities, which are associated with improved physical fitness and mental health outcomes [16]. Because of their many direct and indirect uses, wetlands provide provisioning, regulating, cultural, and supporting ecosystem services, and may generate substantial economic value alongside ecological and social benefits [28, 29, 31].

Furthermore, wetland restoration projects offer opportunities for community engagement and education, fostering a sense of stewardship and awareness about environmental issues among urban residents. In this way, urban wetlands contribute not only to ecological sustainability but also to social and cultural sustainability by supporting community cohesion and environmental literacy [15].

According to Nepal's National Population and Housing Census 2021, as of the census day (25 November 2021), the country's population was 29,164,578, with an average annual population growth rate of 0.92% between 2011 and 2021 [35]. The country has around 27% of its population in urban areas [35], with local governments and development partners increasingly implementing public open space initiatives in secondary cities and municipalities that also aim to maintain key ecological functions, including wetland-related services [10, 11, 19]. In this context, wetland-integrated public open space projects are framed to support accessibility, safety, and inclusion, while contributing to climate resilience at the neighborhood scale [10, 11].

Although local government responses to climate risk often combine disaster management planning, awareness activities, and investments in risk-reduction infrastructure including flood-control works, translating these efforts into inclusive public space outcomes depends on ongoing and sustained resourcing, proper training, and the inclusion of vulnerable groups in planning and maintenance routines [5, 49].

This chapter presents findings from a case study of Kamal Pokhara in Tansen Municipality, situating the analysis within the broader technical assessments and participatory planning and designing processes carried out under the *Cities 4 Women: Inclusive and Climate Resilient Urbanization in Nepal* project. In Nepal, municipalities are implementing public open space projects with the support of development partners, namely, UNOPS, UN-Habitat and Cities Alliance, through the project *Cities 4 Women: Inclusive and Climate Resilient Urbanization in Nepal*, co-funded by the European Union and the Government of Finland, and guided by a steering committee led by Nepal's Ministry of Urban Development [13, 44, 48].

The programme is designed to strengthen women's leadership in municipal planning and budgeting, and to expand their participation in public open space co-design processes through tools that include safety audits, urban walks, and community mapping [13, 41]. In this case study, these tools are applied to a wetland-adjacent public open space, aiming to support accessibility and safety while maintaining ecological functions.

As part of the *Cities 4 Women: Inclusive and Climate Resilient Urbanization in Nepal* project, Nepal has also used an International Urban Design Competition to inspire design proposals for gender-responsive and climate-resilient public open spaces in the project participating municipalities; the competition is not analyzed here as an evaluated intervention outcome, but treated as a mechanism to generate design concepts and suggest implementation pathways for the municipalities [10, 11]. The specific site analyzed in this chapter was co-designed with communities.

The chapter draws on research on public open spaces in Nepal and on data collected through mapping, prioritization, and co-design workshops focused on urban wetlands and adjacent public open spaces. It situates the case within a collaborative planning framework that uses participatory tools (e.g., community mapping, safety audits/urban walks, and co-design workshops) to support the planning and re-design of wetland-adjacent sites.

The chapter explores the following questions:

1. How do infrastructure conditions in Tansen (access, lighting, water condition/quality concerns, and water-edge conditions) shape the use of public open spaces that are adjacent to wetlands?
2. How do institutional arrangements (mandates, budgets, maintenance responsibilities, and water governance) enable or inhibit the delivery of public open spaces that are wetland-sensitive and gender-responsive?

To answer these questions, the Results and Discussion section is organized through analytical filters—infrastructure, institutional, and perceptual—used to examine how physical conditions, governance arrangements, and user experiences interact in shaping wetland-adjacent public open space outcomes in Tansen.

2 Study Area: Urbanization in Nepal

This chapter focuses on Tansen Municipality as a secondary city experiencing rapid urban change, using the wetland-adjacent public open space of Kamal Pokhari as a case to examine how urbanization pressures intersect with climate risk and inclusive public open space delivery. The location of the case site within the municipality is shown in Fig. 1.

Urbanization in Nepal is a rapidly evolving phenomenon, significantly influencing the country's economic, environmental, and social landscapes. Although Nepal has historically been a predominantly rural society, the last few decades have witnessed a steady increase in urbanization, driven by factors such as economic development, rural-to-urban migration, administrative restructuring, and infrastructure expansion [20, 24]). According to Nepal's Degree of Urbanization (DEGURBA) analysis of the 2011 and 2021 census data, approximately 27.07% of Nepal's population resides in urban areas (39.75% in peri-urban and 33.19% in rural), a figure likely to increase as peri-urban areas urbanize and secondary cities expand [20, 34]. Nepal's urban growth is characterized by two major trends: the expansion of existing urban centers such as Kathmandu, Pokhara, and Bharatpur, and the official reclassification of rural municipalities into urban municipalities through administrative reforms. The government's decision to expand the number of urban municipalities substantially during the 2010s has contributed to a rapid statistical increase in urbanization, even though many newly designated urban areas lack essential infrastructure and services [20].

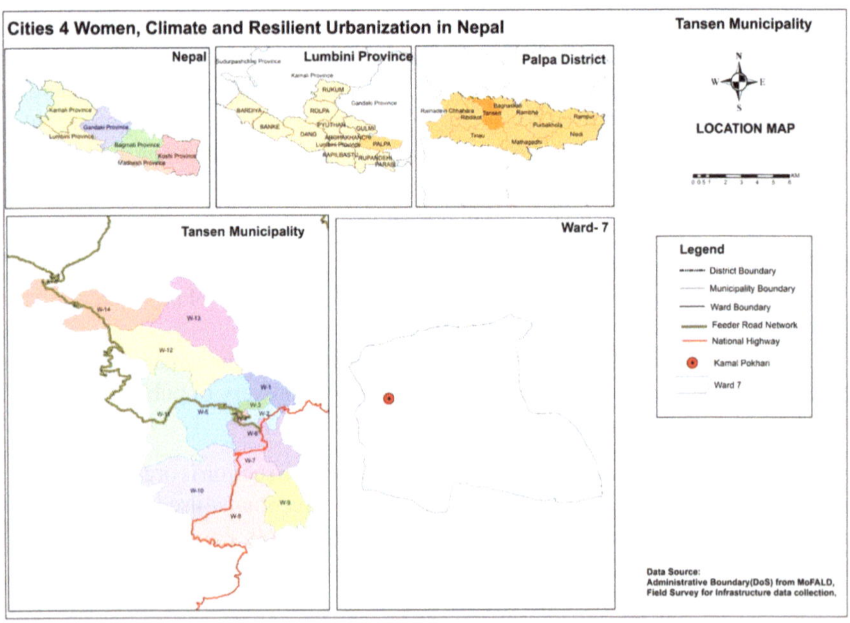

Fig. 1 Location Map of Kamal Pokhari at Tansen Municipality © Authors with Sugam Gautam

One of the defining features of Nepal's urbanization is its uneven spatial distribution. The Kathmandu Valley remains the dominant urban hub, attracting large-scale migration due to better economic opportunities, education, healthcare, and infrastructure [20]. However, this has led to unplanned urban sprawl, traffic congestion, air pollution, and inadequate public services [20]. In contrast, urban municipalities and sub-metropolitan cities such as Birendranagar, Tansen, Sunwal, Hetauda, Chandragiri, Tilottama, and Dhangadhi among many, are experiencing growing urbanization pressures, often without adequate urban planning frameworks [20].

The rapid pace of urbanization in Nepal has created multiple challenges, particularly in infrastructure, environmental sustainability, and governance, with one of the most pressing concerns being the degradation of natural ecosystems, including wetlands, riverbanks, and forests, which are being encroached upon for residential, commercial, and industrial development. Wetlands, in particular, face threats from land reclamation, pollution, and inadequate waste management, reducing their ability to regulate flooding, purify water, and support biodiversity [33]. Additionally, Nepal's cities struggle with climate resilience, as urban expansion increases exposure to floods, landslides, and heat stress. With changing precipitation patterns, many urban centers face water scarcity during the dry season and flooding during monsoons, exacerbated by poor drainage systems and haphazard land-use changes. Urban poverty, inadequate housing, and informal settlements further compound these challenges, leaving marginalized populations highly vulnerable to environmental risks. Despite these challenges, Nepal's urbanization presents opportunities for sustainable development through integrated planning, climate resilience strategies, and nature-based solutions. Green infrastructure, including urban wetlands, parks, and green corridors, offers a pathway to improving urban livability, biodiversity conservation, and climate adaptation [20, 33].

Recent initiatives, such as the *Cities 4 Women: Inclusive and Climate Resilient Urbanization in Nepal* project, emphasize the role of inclusive and gender-responsive public open spaces in fostering urban resilience [11, 48]. By incorporating wetlands into urban planning frameworks, Nepalese cities can enhance ecosystem services, reduce disaster risks, and create healthier, more livable urban environments. Strengthening policy coherence, institutional coordination, and community engagement will be critical to ensuring that urban growth aligns with global sustainability goals such as the 2030 Agenda for Sustainable Development (SDGs) and the New Urban Agenda [21, 45, 46].

2.1 Urbanization and Wetlands in Nepal

The Ramsar Convention on Wetlands, adopted in Ramsar, Iran, in 1971, provides an intergovernmental framework for the conservation and wise use of wetlands. Nepal is a Contracting Party to the Convention, formalizing its commitment to wetland protection [40]. Wetlands, however, have long been overlooked in spite of their significance on a social, economic, religious, and tourism significance [40]. Wetlands in Nepal are vulnerable to both common and region-specific dangers due to the variety of ecological zones in which they are found. Koshi Tappu was designated as Nepal's first Ramsar Site [38]. Across Nepal, ten Ramsar Sites are listed in the RSIS export (Table 1), including four high-altitude lake sites; Gokyo and Associated Lakes, Gosaikunda and Associated Lakes, Phoksundo Lake, and Rara Lake [38].

Local communities manage a large number of wetlands through user groups that support resource conservation and sustainable practices. National policy sources estimate that wetlands cover approximately ~5% of Nepal's land area, encompassing diverse wetland types such as rivers and floodplains, reservoirs, and human-modified wetland landscapes such as paddy fields [33]. By promoting sectors including agriculture, tourism, fish farming, and irrigation, wetlands support local, and national economies [33].

The Ministry of Forests and Soil Conservation (MoFSC) administered wetlands in Nepal through the Department of Forests outside protected areas and the Department of National Parks and Wildlife Conservation (DNPWC) inside protected areas, while

Table 1 List of Ramsar sites in Nepal

No	Listing no	Ramsar sites	District	Designation date	Area (Km^2)	Zone
1	380	Koshi Tappu	Sunsari	17/12/1987	175	Terai
2	1313	Beeshazari and Associated Lakes	Chitwan	13/08/2003	32	Terai
3	1314	Ghodaghodi Lake Area	Kailali	13/08/2003	25.63	Terai
4	1315	Jagadishpur Reservoir	Kapilvastu	13/08/2003	2.25	Terai
5	1692	Gokyo and Associated Lakes	Solukhumbu	23/09/2007	77.7	High Himalaya
6	1693	Gosaikunda and Associated Lakes	Rasuwa	23/09/2007	10.3	High Himalaya
7	1694	Phoksundo Lake	Dolpa	23/09/2007	4.94	High Himalaya
8	1695	Rara Lake	Mugu	23/09/2007	15.83	High Himalaya
9	1850	Mai Pokhari	Ilam	20/10/2008	0.9	Mid-hill
10	2257	Lake Cluster of Pokhara Valley (Phewa, Begnas, Rupa, Dipang, Khaste, Neureni, Maidi, Gunde, & Kamal Pokhari)	Kaski	02/02/2016	261.06	Mid-hill

Source Ramsar sites Information Service [39]

DNPWC served as the Ramsar Administrative Authority [14]. Wetlands conservation and management is integrated into multiple agencies in Nepal, including sectoral ministries and commissions responsible for agriculture, land and irrigation, planning, and water/energy coordination, alongside specialized committees and departments [33]. Newly formed local Governments (Metropolitan, Sub-metropolitan, Municipality, and Rural Municipality) are also investing in wetlands including the Ramsar sites and other local wetlands [33]. Similarly, IUCN Nepal, and the International Center for Integrated Mountain Development (ICIMOD) support wetland conservation and knowledge development through research, technical assistance, and capacity development partnerships [33].

Nepal has developed various master plans and strategies for the conservation and development of its wetlands. These plans focus on biodiversity conservation, flood management, tourism, and sustainable livelihoods for local communities [33]. Guided by the Constitution of Nepal, various strategies and policies support wetland management, including the National Wetland Policy (2003, 2012), the National Ramsar Strategy and Action Plan, Nepal (2018–2024), and the National Urban Development Strategy (2017) [33]. This chapter treats the National Wetland Policy (2012) and the National Ramsar Strategy and Action Plan (2018–2024) as the key national framework references for wetland governance and implementation [33]. The National Wetlands Policy, 2003, emphasizes sustainable wetland management with active community involvement. Scientific study, biodiversity preservation, the protection of endangered species, women's involvement, public awareness, and international obligations are some of its main goals. The policy classifies wetlands for efficient management and aims to stop detrimental activities by regulating damaging behaviors, controlling invasive species, and launching educational efforts [33].

Degradation of grasslands and wetlands is a concern for environmentalists in Nepal and across the world. Ecological equilibrium depends on efficient management. Local initiatives to remove invasive species from the dwindling Sano Narayani water source in the Namuna Buffer Zone Community Forest serve as an example of effective conservation. Researchers identified major risks to wetland and grassland ecosystems, including siltation, encroachment, resource misuse, invasive species, pollution, overgrazing, and improper burning [6].

The *Cities 4 Women: Inclusive and Climate Resilient Urbanization in Nepal* Project supports municipal public open space planning and delivery in Nepal. Its four sub-goals are to increase demand for climate-resilient, gender-responsive public open spaces, create participatory public open space projects, strengthen municipal urban planning and design capacities, and assess the legal framework in order to suggest policy changes. Some of the public open spaces that were designed or redesigned through participatory approach were wetland-adjacent spaces. Although public open space is increasingly recognized in Nepal's urban development practices, planning for purposes of climate resilience and urban health still has a way to go. Recognizing this gap, this chapter discusses public open spaces that are adjacent to wetlands, examining how participatory, gender-responsive processes can connect the needs of daily public-realm with longer-term blue-green objectives of resilience.

3 Materials and Method

3.1 Case Study Design and Analytical Framework

This chapter uses a single-case study design, focusing on one wetland-adjacent public open space in Tansen Municipality: Kamal Pokhari. The analytical framework applies three filters—infrastructure, institutional and perceptual—to examine how blue-green infrastructure conditions and governance shape human well-being and equity outcomes in urban systems [2, 3].

3.2 Case Study Selection

The site was selected for three reasons. First, Kamal Pokhari directly interfaces with a wetland/water-body condition (an urban pond). Second, it is actively used as public open space. Third, it was included in the *Cities 4 Women: Inclusive and Climate Resilient Urbanization in Nepal* project participatory process during the study period.

The selection of Kamal Pokhari followed the ward-level participatory process used by local governments in Nepal. A Participatory Community Mapping Workshop was organized by the project in Ward 7, Tansen Municipality, on 30 May 2024, and facilitated as an interactive session on participatory community mapping aligned and informed by the local government's seven-step planning process, which begins with ward-level consultations to gather community input.

The workshop was attended by 30 participants, including ward members and representatives from Neighborhood Development Committees/Organizations (Tole Bikash Sanstha). Twenty-three participants were women and seven were men. The workshop included interactive discussions with various community groups, ward members, and other stakeholders.

From the workshop, 24 public open spaces were mapped. Following discussion and prioritization, three public open spaces were prioritized: Bansha Gopal park, Ranikohal Khel Maidan, and Kamal Pokhari. The workshop minutes were recorded, and signed at the end of the session. Kamal Pokhari was selected as the case site for this chapter (see Fig. 2).

Kamal Pokhari is a public open space locally recognized for its connections to culture and place-identity, while the pond is described by participants as one of Tansen Municipality's artificial lakes, spatially linked to the nearby Bansha Gopal area, which holds religious and historical significance within the municipality.

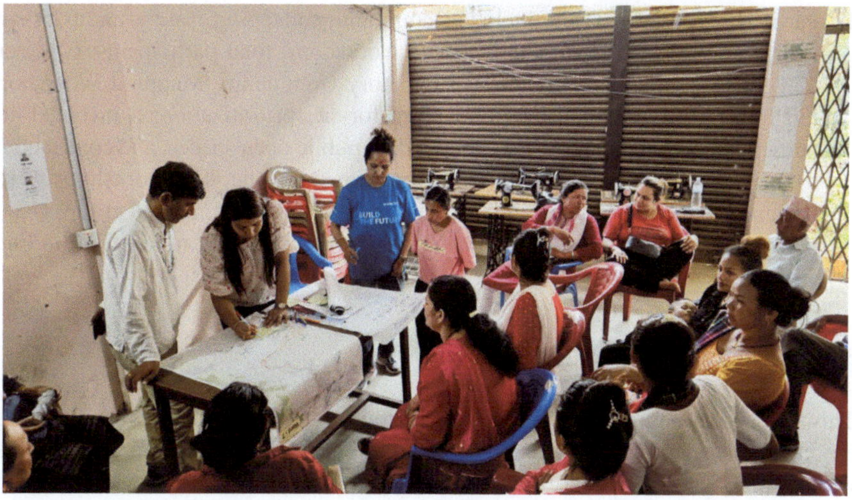

Fig. 2 Community Mapping Workshop at ward no.7 © Authors

3.3 Data Sources and Collection

The study utilized a mixed-methods approach, combining (i) document and policy review, (ii) a structured user/community survey at the case site, (iii) key-informant interviews, and (iv) participatory tools implemented through the *Cities 4 Women: Inclusive and Climate Resilient Urbanization in Nepal* process. Data collection was conducted within the project context; however, the site-based analysis in this chapter focuses on a single case in Tansen Municipality (Kamal Pokhari).

1. Document and policy review. These included municipal plans, relevant national legal and policy instruments (e.g., Local Government Operation Act [LGOA] 2017), and program documentation. In addition, the review drew on relevant government planning/budgeting guidance and constitutional/fiscal instruments referenced in the programme documentation, as well as the 2021 census report where relevant for contextual indicators.
2. Structured user/community survey. This was administered at the site and the area surrounding the case study site (Kamal Pokhari) to capture users' experiences and priorities related to access, safety (including lighting), and wetland-adjacent conditions. Given the study's applied, site-based focus, the survey is treated as a descriptive intercept survey of users present during field visits (i.e., approached on-site and invited to participate voluntarily).
3. Key-informant interviews. This consisted of semi-structured interviews with municipal officials, ward representatives, community leaders, and relevant agencies, to clarify mandates, budgeting and implementation constraints, and responsibilities for maintenance and water governance relevant to wetland-adjacent public open space delivery.

4. Participatory tools: These included community mapping, safety audits/urban walks, and co-design workshops (three workshops; total participants: 67). For Tansen Municipality, a Ward 7 Participatory Community Mapping Workshop (30 May 2024) provided the entry point for site prioritization; it involved 40 participants (23 women; 7 men), mapped 24 public open spaces, and contributed to the selection of Kamal Pokhari as the case study site. The workshops were designed to include women and girls, as well as other marginalized groups (e.g., persons with disabilities, senior citizens, sexual, and gender minorities), in line with project inclusion objectives.

3.3.1 Programme-Wide Data Context

This subsection situates the Tansen case within the wider project while keeping the chapter's empirical analysis focused on Kamal Pokhari. The purpose of this wide framing is to situate the Tansen case within the wider initiative, while keeping the chapter's empirical analysis focused on the single case study site described in Sect. 3.1.

In the *Cities 4 Women: Inclusive and Climate Resilient Urbanization in Nepal* project, data collection and participatory tools were applied across participating municipalities to support public open space planning with a focus on social inclusion and gender responsiveness. Programme tools included gender-sensitive urban climate and public space assessment approaches, community mapping, and safety audits/urban walks, designed to bring forth urban experiences and identify priorities for co-design. Remaining in line with the programme's objectives, participation processes sought to include women and girls and to engage a diversity of residents, including disadvantaged and under-represented groups, to better understand differentiated needs and barriers in public open space use.

For this chapter, both primary and secondary data sources were used. Secondary sources included Nepal's census reporting for contextual indicators, together with relevant national and local planning and governance instruments referenced in the programme documentation such as the Local Governance Operation Act (LGOA), 2017, and associated planning/budget guidance, where applicable. Primary sources included participatory workshops and associated outputs including mapping and prioritization materials, consultation meetings, and semi-structured key-informant interviews with relevant municipal/ward actors, and a structured user/community survey conducted at the case site.

3.4 Analysis

The qualitative material collected from the research (semi-structured interviews, workshop outputs, and field notes) was coded thematically using a hybrid deductive-inductive approach. Deductive coding followed the chapter's analytical framework and its three filters, (i) infrastructure, (ii) institutional, and (iii) perceptual, while inductive codes were developed iteratively to capture themes emerging from the case material. Coding was conducted using a single-coder approach with an explicit codebook and memoing to maintain an audit trail; ambiguities were resolved through iterative review of raw notes and workshop documentation. Survey data were analyzed using descriptive statistics. Document review findings were synthesized to describe governance and planning context, and were used to contextualize and cross-check claims emerging from interviews and participatory outputs.

3.5 Participatory Tools and Co-Design Workshop

Participatory procedures are widely used in urban development to strengthen inclusion in planning and design. In this chapter, participation is treated not as symbolic consultation but as a mechanism for sharing knowledge, negotiating priorities, and shaping implementable public space proposals with communities and institutions. This approach is consistent with Arnstein's [4] argument that meaningful participation occurs when communities are able to influence decisions rather than being limited to symbolic consultation. Therefore, within Arnstein's "Ladder of Citizen Participation," the concept of Partnership sits within the broader category of "degrees of citizen power" [4], implying shared deliberation and negotiation between communities and institutional actors.

The project is framed in a similar manner, creating repeated structured opportunities for women and other under-represented/marginalized users to voice their needs, identify use barriers, and collaborate in developing priorities that municipal and ward-level actors can take forward in planning, budgeting, and implementation. While this does not yield full citizen control, it goes beyond consultation.

The process began with Community Mapping: the activity was held by localizing the Cities Alliance tool "Women Engagement in Cities" where public from the community, technical person from executive office and local leadership were brought together to map their respective community. Participants included local community members, the Mayor, Executive Members, the ward chairperson and members, representatives from *Tole bikash Samiti* (TLO), mothers' groups, senior citizens, etc. The understanding is that the local people are aware of their locality more than any other expert. The process focused on (i) identifying the public open spaces used most frequently, (ii) discussing patterns of use by time and user group, and (iii) prioritizing spaces using criteria that included cultural and recreational value as well as safety, security, and accessibility. The process followed six steps:

- Step 1: Two-hour collaborative workshops were held, bringing together approximately 40 community members under the leadership of the Ward Chairperson.
- Step 2: Participants were introduced to the project (new to the ward), oriented to the purpose of the event, and were invited to discuss which public open spaces they use most frequently, at what time, and for what purposes.

- Step 3: Participants discussed in small groups to identify and list all public open spaces (POS) in the ward, recording them on a shared chart.
- Step 4: The identified POS were called out, and participants discussed how each space affects their daily life, including patterns of use and key constraints.
- Step 5: Participants prioritized POS based on cultural and recreational activities, and discussed safety, security, and accessibility concerns across gender and age groups, including persons with disabilities.
- Step 6: The prioritized POS were then discussed in more detail to clarify why and how each site was considered a priority, and to identify site-specific problems and potential responses aligned with the project objectives.

A key element of this stage was gender-based mapping, used to document how women and other users experience public open space differently, identifying popular places of importance (e.g., sacred/cultural sites, historical places, populated streets) and less popular ones, and how each is associated with perceived risks and constraints. During the process, participants annotated maps with the location of public open space, while also reporting concerns relating to inadequate lighting, visibility, access routes, and barriers for persons with disabilities (e.g., lack of step-free routes and safe surfaces, lack of accessible toilets), and locations perceived as unsafe. The purpose of the exercise was to make these everyday experiences legible to decision-makers, grounding subsequent design work in the priorities of the users' reports rather than in generic assumptions about the needs of the public open space. In the Tansen case, these concerns informed prioritization and co-design discussions, especially regarding night-time visibility, perceived harassment risk, and the accessibility of paths and facilities for different age groups and abilities.

Following Community Mapping, Co-Design Workshops were used to translate the prioritized needs into practical design directions for the selected site. The workshop combined facilitated discussion (first session), a site walk/safety audit (second session), and participatory prototyping (third session) to produce user-led design priorities and spatial ideas. In the facilitated discussion (first session), participants identified priority themes related to inclusion, accessibility, safety, and everyday usability, while the facilitator took notes. During the site walk (second session), participants engaged in mind-mapping for design, after assessing the wetland interface and the conditions of the surrounding public realm (i.e., road accessibility, circulation routes, edge conditions, visibility, lighting), so that the design ideas remained grounded in on-site constraints and opportunities. Then, in the prototyping phase (third session), participants were asked to visualize and articulate how they would like the public open space to function and feel in everyday use. To support inclusive discussion, the workshop participants were divided into two groups, ensuring that one group included women, children, senior citizens, and persons with disabilities, so that accessibility and safety needs were explicitly considered in the concept development. Each group then produced a simple prototype using sketches, diagrams, or models, expressing its design priorities. Feedback from both groups was incorporated into a shared concept, enabling the project team to understand how residents perceived and prioritized the co-created ideas.

In Ward 7, Tansen Municipality, the co-design workshop focused on the prioritized public open space of Kamal Pokhari (Fig. 3), where participants discussed how the pond's wetland interface and surrounding public realm could be improved to support everyday use while maintaining ecological functions. The resulting priorities were synthesized into site diagrams identifying the wetland interface, access structure, and priority co-design areas (Fig. 4). Community proposals emphasized culturally distinctive design motifs, including references to designated local products such as *karuwa* and *dhaka*, which informed ideas for an entrance feature and photo point/selfie spot.

The co-design process also raised forward site-management issues, such as the allocation and management of a parking area near the entrance, the provision of shaded seating around the water body, and the location of basic services including inclusive amenities, accessible toilets ("smart toilet") and safe drinking-water provision (water dispenser), proposing that circulation routes should be made more inclusive and accessible through tactile guidance and no-slip paving. Additional proposals related to enhancing the wetland-edge experience and improving safety and legibility, including a viewing deck, seating, lighting, signage, CCTV, and the display of emergency contact information. Furthermore, participants proposed leisure and identity features such as a multi-color fountain concept that would retain lotus and enhance lake views, and a small secretariat/office space to support maintenance and control. Linkages to nearby heritage and religious destinations were also discussed, including a proposed fly-over connection toward Surya Kunda and associated municipal plans (e.g., concept described as "108 taps"), as well as the concept of an open library near the children's park (Fig. 4).

Overall, the participatory process translated user experiences into context-specific design proposals, strengthening dialog among community participants, ward representatives, and municipal technical staff, and increased the participation of women, girls, and other marginalized communities in urban planning and design processes. In the Tansen case, the process brought forward interventions that deal with everyday inclusion of accessibility, safety, comfort, and legibility, while keeping focus on wetland interface as a blue-green asset. This combination is presented as a basis for improving local ownership, future implementation readiness, and longer-term stewardship of wetland-adjacent public open spaces.

Fig. 3 Co-design workshop at Kamal pokhari Tansen municipality © Authors

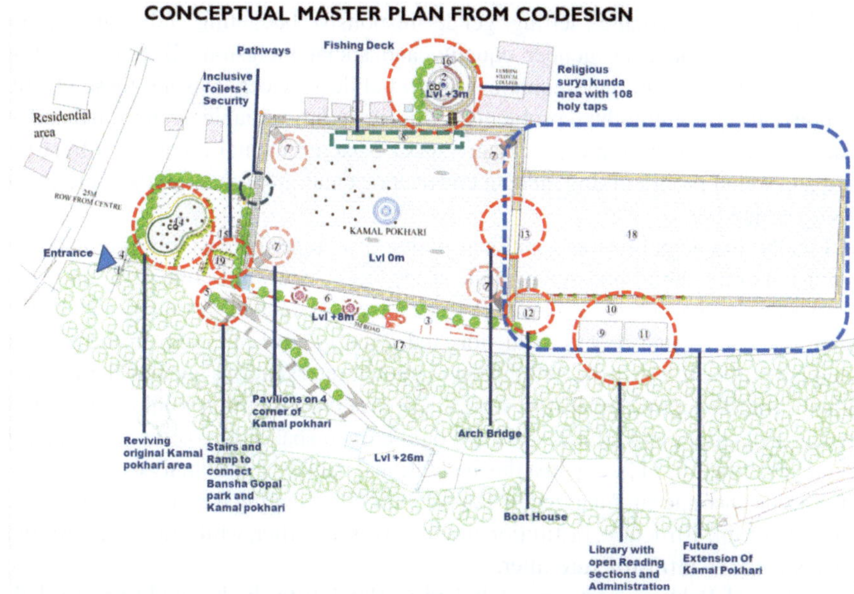

Fig. 4 Site diagrams (Kamal Pokhari): wetland interface, access structure, and priority co-design areas © Authors

4 Results and Discussion

4.1 Institutional Context: Cities 4 Women and Municipal Implementation Conditions

The *Cities 4 Women: Inclusive and Climate Resilient Urbanization in Nepal* project provides the institutional platform through participatory tools, such as community mapping, safety audits/urban walks, and co-design workshops, were used to identify priorities and translate them into preliminary design directions for public open spaces adjacent to wetlands in Tansen Municipality. Therefore, the results presented here reflect not only physical needs and site conditions, but also the interpretations and priorities expressed by participants during the process, as well as the delivery constraints by the municipality.

Across the partner municipalities involved in the project, municipal planning was described by stakeholders as a combination of public consultations, ward-level prioritization, and budget allocation in an effort to create alignment with national budgeting procedures. Despite these municipal planning processes, interviewees noted recurring constraints in implementation, including the lack of consistent involvement of the community through representative structures (e.g., Tole-level groups), delays with regard to the execution of the project, and limited room for decision-making when higher-level guidelines dictate how funds should be spent.

Within this governance setting, gender-responsive budgeting and related initiatives sought to advance gender equality and social inclusion (GESI), intending to expand participation and equal access to public services. Despite these efforts, respondents also highlighted problems to this, including variable institutional understanding of GESI principles, a disagreement between capital-intensive infrastructure and social programming, and an underrepresentation of marginalized groups in decision-making.

Finally, municipalities and national policy discourse increasingly reference the provision of neighborhood-scale public space (e.g., "one ward - one park" and "one ward - one playground") alongside broader agendas related to climate-resilience, such as those related to risk reduction and flood control [19, 43]. In the project, such priorities are also reflected through targeted interventions such as the International Urban Design Competition, which called for proposals that would demonstrate gender-responsive and climate-resilient public open spaces across the participating municipalities [10, 11]). Against this institutional backdrop, the analysis proceeds to the Tansen case site summarized in Table 2, before presenting the findings structured through the infrastructure, institutional, and perceptual filter, while Table 3 represents results with the infrastructure filter.

At Kamal Pokhari, participants linked wetland/water-body conditions to daily comfort and safety, including shade, microclimate, views, and recreation, while also identifying barriers to universal access, insufficient lighting, poor path continuity, and

Table 2 Case study snapshot (Tansen Municipality)

Site name	Wetland/ water-body interface	Dominant users/ times	Key risks	Priority co-design interventions	Responsible institutions for maintenance/ water governance
Kamal Pokhari	Central urban pond with landscaped edges but limited physical water access	1. Walkers and joggers (morning/ evening), 2. Families and youth (afternoons), elderly users (morning), visitors (mostly on weekends)	1. Water quality deterioration 2. Safety concerns for children and elderly 3. Conflicts between conservation and recreation 4. Encroachment	1. Inclusive edge design with safe, stepped access zones 2. Shaded seating 3.Community monitoring of water quality 4. lighting and visibility 5. Clear zoning for ecology, leisure, and circulation	1. Tansen Municipality 2. Ward office 3. Community user groups/ local committees

Source © Authors

Table 3 Infrastructure filter: physical access, safety, and wetland condition

Aspects	Findings/issues	Survey results	Participant quotes	Observations
Access and paths	Barriers to universal access; discontinuous paths; uneven surfaces	84% reported difficulty accessing certain paths or areas due to uneven surfaces and missing ramps	**"The path near the lotus pond is broken; I cannot walk there safely."**	1. Sections of the pathways were interrupted as it was not continued throughout the water body 2. Vegetation overgrowth
Lighting and Safety	Insufficient lighting; concerns about night-time safety	88% reported poorly lit areas	**"It's dark near the water edge; I don't feel safe walking there after sunset."**	Lighting poles absent or non-functional in key zones
Wetland/ Water-edge Condition	Safety and water quality concerns; limited recreational usability	76% concerned about water edge safety	**"The water near the edge looks dirty; after lotus flower blooming time is completed the decayed flower degrades the water quality."**	Water edge exposed, uneven; some areas prone to littering or erosion

concern about water quality and edge safety. Together, Tables 2 and 3 summarize the site, locating the issues raised by participants, including access, lighting, water-edge condition, and perceived safety. These are useful to go through prior to reviewing the site-specific evidence. Across the investigated site and the main issues raised by participants before the discussion turns to the site-specific evidence.

4.2 Institutional Filter: Mandates, Budgets, and Maintenance/Water Governance

In the cases of Kamal Pokhari, the participants' design priorities, which included lighting, paths, toilets, wayfinding, and water-edge measures, are closely related to the institutional question of who funds, builds, and maintains them over time. The discussions with the stakeholders highlighted that delivering public open spaces situated adjacent to wetlands requires the cross-coordination of different sectors (public works, environment/forestry, ward offices, and any lake/wetland committees). Commonly identified gaps included unclear maintenance responsibility for lighting and paths, limited budget continuity at the end of project cycles and beyond, and weak prevention mechanisms against encroachment and dumping.

Table 4 The main user-group concerns and the supporting evidence

User groups	Key perceptions and concerns	Evidence	Evidence type
Women	Safety at night, harassment risk	"I avoid the pond after dark because the path is poorly lit."	Survey/quotes
Old age people	Lack of seating area; poor path accessibility	Uneven pathways make movement throughout the site more difficult	Observations
Adolescents	Need for spaces to socialize; freedom of movement	"There aren't any areas where we can gather."	Survey/quotes
Person with Disabilities	Ramp access; toilet accessibility	Absence of appropriate ramps and toilet facilities	Observations
Children	Need for suitable play areas	Lack of child-friendly equipments in the play area	Observations

4.3 Perceptual Filter: Cultural Value, Belonging, and Gendered Safety

The perceptual filter highlights how different social groups experience Kamal Pokhari in uneven ways, especially in relation to safety, dignity, comfort, and ease of movement. These perceptions matter because they shape whether the site is understood as welcoming and usable in everyday life. Table 4 summarizes the main user-group concerns and the supporting evidence from surveys, quotations, and observations. The outputs of the workshop suggest that the perceptions of the users' perceptions combined cultural and religious meanings with practical concerns regarding harassment risk, night-time visibility, and the adequacy of toilets, seating, and wayfinding. Workshop discussions and interviews raised concerns on the experiences of safety, dignity, and access, as they differed across social groups, including caste-oppressed communities and sexual and gender minorities. Some participants linked perceptions of inclusion to whether facilities and design features support diverse needs, such as safe lighting, toilets, seating, and privacy considerations.

4.4 Synthesis: How the Three Filters Interact Across the Site

Across the site of Kamal Pokhari, the three filters—infrastructure, institutional, perceptual—interacted in meaningful ways. Infrastructure conditions, including discontinuous access, lighting gaps, and water-edge safety and quality concerns, defined patterns of use and perceptions of comfort and security. Institutional arrangements, such as mandates, budget continuity, and maintenance and water governance responsibilities, affected the possibility and potential of implementing proposed upgrades that would also be sustainable beyond project cycles. In turn, these constraints and enabling factors influenced perceptions of belonging, cultural value, and gendered safety, which were repeatedly reflected upon during the participatory process of co-designing. Translating this synthesis through the blue-green infrastructure lens, the cases suggest that realizing wetland-related benefits for people's well-being and equity depends on whether the site's design implements durable maintenance and stewardship routines, rather than creating separations between the acts of wetland protection and public-realm upgrades [2, 3].

5 Conclusion

Urban wetlands in Nepal provide critical ecosystem services by regulating hydrological cycles, mitigating flood risks, support biodiversity, and sustaining cultural practices and everyday social uses such as gathering and recreation. Nepal has established frameworks to support wetland conservation, yet, rapid urban development, encroachment, waste disposal, agricultural runoff, fragmented governance, limited funding, and weak implementation continue to undermine their effectiveness. Integrating wetlands into urban planning offers significant opportunities for climate mitigation and adaptation, while their economic potential is in ecotourism and recreational activities provide avenues for sustainable economic growth.

The case studies of Nepal illustrate the transformative potential of wetlands when integrated into urban planning frameworks. Future efforts should focus on enhancing institutional capacities, fostering multi-stakeholder collaboration, and leveraging innovative financing mechanisms. By aligning wetland conservation with global development agendas, cities can enhance resilience, equity, and sustainability, paving the way for a coexistence between urban development and natural ecosystems.

In this chapter, the design ideas emerged through structured community engagement, which strengthened local ownership and improved implementation readiness. Feasible, inclusive and accessible venue selection creates a favorable and credible environment for discussion, together with the participation of municipal leadership during workshops and site visits, helped build trust in the process and deepen the collective understanding of the site's environmental, historical, religious, cultural, social and land ownership aspects.

In Kamal Pokhari, the co-design process showed that diverse participation, such as community members, ward leadership, and municipal technical staff, accelerated the sharing of information and understanding of the sites' constraints and opportunities, including ecological trade-offs at the water edge and the daily usability requirements of different groups. From the participants' standpoint, priorities included interventions that combined stewardship sensitive to wetlands, such as protecting water bodies and edges, managing runoff, and strengthening planting buffers. Priorities further included the upgrading of public spaces to support inclusion, such as having continuous accessible paths, safer lighting, seating, toilets, and legible wayfinding. Equally contributing to the study's outcomes was the presence of municipal leadership during site walks and workshops. This participatory presence helped align community preferences with feasible implementation ideas, reducing the likelihood of producing proposals that could potentially damage the wetlands' functions.

These findings reinforce the chapter's core argument; that embedding wetland-sensitive public open space design into municipal planning and maintenance routines, paired with ongoing community stewardship and clearer cross-agency coordination, can deliver simultaneous gains of equity, safety, and climate resilience.

Authors' Contribution First draft of this chapter was written by Ilija Gubić with contribution by Anup Ghimire to the introduction and study areas sub-chapters, while second draft was written by Ilija Gubić and Stefania Leontiadou with contribution from Manju Raut on case-study information.

Funding This research received no funding.

Data Availability Statement The datasets are available from the corresponding author on reasonable request.

Potential Conflicts of Interest On behalf of the authors, the corresponding author states that there is no conflict of interest. At least one of the co-authors participated in each of the activities mentioned in the chapter that were part of the *Cities 4 Women: Inclusive and Climate Resilient Urbanization in Nepal* project implemented by UNOPS, UN-Habitat and Cities Alliance, funded by European Union and Finland, and led by the Ministry of Urban Development.

Informed Consent Informed consent was obtained from all individual participants included in the study.

References

1. Alikhani S, Nummi P, Ojala A (2021) Urban wetlands: a review on ecological and cultural values. Water 13 (22):3301. https://www.mdpi.com/2073-4441/13/22/3301
2. Andersson E, Borgström S, Haase D, Langemeyer J, Wolff M, McPhearson T (2021) Urban resilience thinking in practice: ensuring flows of benefit from green and blue infrastructure. Ecol Soc 26(4):39. https://doi.org/10.5751/ES-12691-260439
3. Andersson E, Langemeyer J, Borgström S, McPhearson T, Haase D, Kronenberg J, Barton DN, Davis M, Naumann S, Röschel L, Baró F (2019) Enabling green and blue infrastructure to improve contributions to human well-being and equity in urban systems. Bioscience 69(7):566–574. https://doi.org/10.1093/biosci/biz058
4. Arnstein SR (1969) A ladder of citizen participation. J Am Inst Plann 35(4):216–224. https://doi.org/10.1080/01944366908977225
5. Asian Development Bank (2019) Nepal: flood risk management sector assessment (consultant's report). Asian Development Bank, Manila
6. Bhattarai R (2013) Local communities work to save their grasslands and wetlands. WWF Nepal, 14 Jun. https://www.wwfnepal.org/?209060/Local-Communities-Work-to-Save-their-Grasslands-and-Wetlands
7. Bratman GN, Hamilton JP, Daily GC (2015) The impacts of nature experience on human cognitive function and mental health. Sci Rep 5:10185. https://doi.org/10.1038/srep10185

8. CGED Nepal (2022) Development of management plan of Jagadishpur Bird Sanctuary. Kapilvastu, Lumbini Province. 31 Jul. From CGED Nepal: https://www.cgednepal.org/project/development-of-management-plan-of-jagadishpur-bird-sanctuary-kapilvastu-lumbini-province/
9. Cheshmehzangi A (2020) 10 Adaptive measures for public places to face the COVID 19 pandemic outbreak. City and Society. https://doi.org/10.1111/ciso.12282
10. Cities Alliance (2024) International urban design competition call (document). Cities Alliance. https://www.citiesalliance.org/sites/default/files/2024-08/international_urban_design_competition_call_august_1_2024.pdf
11. Cities Alliance (2025) Cities4Women Nepal: international urban design competition. Cities Alliance. https://www.citiesalliance.org/sites/default/files/2025-06/cities4women_nepal_-_urban_design_competition_2025_application_guidance.pdf
12. Cities Alliance (nd) Cities 4 Women Nepal—overview. Cities Alliance. https://www.citiesalliance.org/how-we-work/country-and-regional-initiatives/cities-4-women-nepal/overview
13. Cities Alliance (2026) Cities 4 Women: integrating gender-sensitive and climate-resilient urban planning for sustainable cities. Cities Alliance. https://www.citiesalliance.org/how-we-work/country-and-regional-initiatives/cities-4-women-nepal/overview
14. DoF (2017) Wetlands of Western Nepal: a brief profile of selected lakes. Department of Forest, Kathmandu
15. Folke C, Hahn T, Olsson P, Norberg J (2005). Adaptive governance of social-ecological systems. Annu Rev Environ Resour 15(30):441–73. https://doi.org/10.1146/annurev.energy.30.050504.144511
16. Gascon M, Triguero-Mas M, Martínez D, Dadvand P, Forns J, Plasència A, Nieuwenhuijsen MJ (2017) Mental health benefits of long-term exposure to residential green and blue spaces: a systematic review. Int J Environ Res Public Health 14(8):940. https://doi.org/10.3390/ijerph14040444
17. Ghermandi A, van den Bergh JCJM, Brander LM, de Groot HLF, Nunes PALD (2010) The values of natural and human-made wetlands: a meta-analysis. Water Resour Res 46(12):1–13. https://doi.org/10.1029/2010WR009071
18. Gill SE, Handley JF, Ennos AR, Pauleit S (2007) Adapting cities for climate change: the role of the green infrastructure. Built Environ 33(1):115–133. https://doi.org/10.2148/benv.33.1.115
19. Government of Nepal, Ministry of Federal Affairs and Local Development (2013) Environment-friendly Local Governance Framework (EFLG). Government of Nepal, Kathmandu. https://www.dpnet.org.np/public/uploads/files/Environment-friendly%202019-04-04%2007-29-49.pdf
20. Government of Nepal, Ministry of Urban Development (2017) National Urban Development Strategy (NUDS), 2017. Ministry of Urban Development, Kathmandu. https://sar-climate.adpc.net/wp-content/uploads/2022/06/NP-RTI-GD-162.pdf
21. Gubic I, Baloi O (2019) Implementing the New Urban Agenda in Rwanda: Nation-Wide Public Space Initiatives. Urban Plan 4(2):223–236. https://doi.org/10.17645/up.v4i2.2005
22. Gubić I, Baloi O (2020) Public open space initiatives for healthier cities in Rwanda. J Public Space 5(2):129–146. https://doi.org/10.32891/jps.v5i2.1287
23. Gubić I, Wolff M (2022) Use and design of public green spaces in Serbian cities during the COVID-19 pandemic. Habitat Int 128:102651. https://doi.org/10.1016/j.habitatint.2022.102651
24. Gubić I, Uprety S, Danilović Hristić, Leontiadis S (eds) 2024 Urban Bootcamp: Public open spaces. Book of abstracts 1st International Conference on Public Open Spaces. Kathmandu: Ministry of Urban Development, European Union, UNOPS, UN-Habitat and Cities Alliance. ISBN: 9789937188937
25. IUCN (2020) Global standard for nature-based solutions: a user-friendly framework for the verification, design and scaling up of NbS (1st ed). International Union for Conservation of Nature, Gland. https://doi.org/10.2305/IUCN.CH.2020.08.en
26. Kennedy C, Pincetl S, Bunje P (2019) The study of urban metabolism and its applications to urban planning and design. Environ Pollut 159(8–9):1965–1973. https://doi.org/10.1016/j.envpol.2010.10.022

27. Kabisch N, Qureshi S, Haase D (2015) Human-environment interactions in urban green spaces - a systematic review of contemporary issues and prospects for future research. Environ Impact Assess Rev 50:25–34. https://doi.org/10.1016/j.eiar.2014.08.007
28. Lannas S K, Turpie J (2009) Valuing the Provisioning Services of Wetlands: contrasting a rural wetland in Lesotho with a peri-urban wetland in South Africa. Ecol Soc 14(2). https://www.ecologyandsociety.org/vol14/iss2/art18/
29. Matthews T, Lo AY, Byrne JA (2015) Reconceptualizing green infrastructure for climate change adaptation: barriers to adoption and drivers for uptake by spatial planners. Landsc Urban Plan 138:155–163. https://doi.org/10.1016/j.landurbplan.2015.02.010
30. McInnes RJ (2014) Recognising wetland ecosystem services within urban case studies. Mar Freshw Res 65(7):575–588. https://doi.org/10.1071/MF13006
31. MEA (Millennium Ecosystem Assessment) (2005) Ecosystems and human well-being: wetlands and water synthesis. World Resources Institute, Washington, DC
32. Mitchell R, Popham F (2008) Effect of exposure to natural environment on health inequalities: an observational population study. Lancet 372(9650):1655–1660. https://doi.org/10.1016/S0140-6736(08)61689-X
33. MoFE (2018) National Ramsar Strategy and Action Plan, Nepal (2018–2024). Government of Nepal, Ministry of Forests and Environment, Kathmandu
34. National Statistics Office (2024) Launching of the report on Degree of Urbanization in Nepal: Press Release (2 January 2024). National Statistics Office, Kathmandu. https://giwmscdnone.gov.np/media/app/public/36/posts/1704183209_71.pdf
35. National Statistics Office (NSO) (2024) National Population and Housing Census 2021: Population Composition of Nepal. National Statistics Office, Kathmandu. https://giwmscdnone.gov.np/media/app/public/36/posts/1710821571_26.pdf.
36. Ouedraogo P (2010, Dec) Rwanda: a case study on wetland governance and its impacts on urban city planning. From The Convention on Wetlands, https://www.ramsar.org/
37. Ramsar Convention on Wetlands Secretariat (2018) Wetlands: essential for a sustainable urban future (Fact Sheet 10). Ramsar Convention Secretariat, Gland. https://www.ramsar.org/sites/default/files/urban_wetlands_en.pdf
38. Ramsar Convention Secretariat (nd) Annotated list of Wetlands of International Importance: Nepal (RSIS export). Ramsar Sites Information Service (RSIS). https://rsis.ramsar.org/sites/default/files/rsiswp_search/exports/Ramsar-Sites-annotated-summary-Nepal.pdf (Accessed 4 Jan 2026)
39. Ramsar Sites Information Service (nd) Explore sites: Nepal (Region: Asia). https://rsis.ramsar.org/?f%5B0%5D=region_en_s%3AAsia&f%5B1%5D=country_en_s%3ANepal (Accessed 4 Jan 2026)
40. Ramsar Sites in Nepal (nd) Ramsar sites in Nepal. National Lake Conservation Development Committee (NLCDC): https://nepallake.gov.np/ramsar-sites/#:~:text=On%20April%2017%2C%201988%2C%20Nepal,%2C%20religious%2C%20and%20touristic%20significance
41. Shrestha B, Gubić I (2025) From data gaps to people powered cities. UNOPS, Copenhagen. Available online: https://www.unops.org/news-and-stories/insights/from-data-gaps-to-people-powered-cities
42. Tafahomi R, Nadi R (2021) Protection of Natural Wetlands through Landscape Design in Kigali City. Rwanda J Eng Sci Technol Environ:1–16
43. TRN Online (2022) Bharatpur metropolis unveils policy, programmes: overhead bridges to be built, electric buses to be used. The Rising Nepal, 24 Jun. https://risingnepaldaily.com/news/13011
44. UN-Habitat Nepal (2026) Cities 4 Women: Inclusive and climate resilient urbanization in Nepal. https://unhabitat.org.np/project_detail/cities-4-women
45. UN General Assembly (2015) Transforming our world: the 2030 Agenda for Sustainable Development (Resolution A/RES/70/1). United Nations, New York. https://www.un.org/en/development/desa/population/migration/generalassembly/docs/globalcompact/A_RES_70_1_E.pdf
46. UN General Assembly (2016) New Urban Agenda (Resolution A/RES/71/256). United Nations, New York. https://www.un.org/en/development/desa/population/migration/generalassembly/docs/globalcompact/A_RES_71_256.pdf

47. United Nations, Department of Economic and Social Affairs, Population Division (2018) World Urbanization Prospects 2018: highlights. United Nations, New York. https://population.un.org/wup/assets/WUP2018-Highlights.pdf
48. UNOPS (2025a) Cities by her, for her. United Nations Office for Project Services (UNOPS), 7 Mar. https://www.unops.org/news-and-stories/stories/cities-by-her-for-her
49. UNOPS (2025b) From data gaps to people-powered cities. UNOPS Insights, 31 October. United Nations Office for Project Services (UNOPS), https://www.unops.org/news-and-stories/insights/from-data-gaps-to-people-powered-cities
50. WHO Regional Office for Europe (2016) Urban green spaces and health. World Health Organization Regional Office for Europe, Copenhagen. https://www.who.int/europe/publications/i/item/WHO-EURO-2016-3352-43111-60341
51. Wolff M, Haase D (2019) Mediating sustainability and liveability—Turning points of green space supply in European cities. Front Environ Sci 7. https://doi.org/10.3389/fenvs.2019.00061
52. World Bank (2021) A catalogue of nature-based solutions for urban resilience. World Bank, Washington, DC. https://documents1.worldbank.org/curated/en/502101636360985715/
53. Zedler JB, Kercher S (2005) Wetland resources: status, trends, ecosystem services, and restorability. Annu Rev Environ Resour 30(1):39–74. https://doi.org/10.1146/annurev.energy.30.050504.144248

From Principles to Design: Sustainable Urban Drainage with Decentralized and Nature-Based Solutions in Developing Economies

Md Sahidul Islam, Bo Sun, Jia-Qiang Lv, and Yanchen Liu

1 Introduction

Urban drainage systems (UDSs) play a key role in safeguarding the public health, economic vitality, and social welfare of a society [7, 19, 22]. However, emerging economies still suffer from inadequate infrastructure. For example, in Dhaka, Mumbai, Jakarta, and Karachi, urban flooding has become one of the most pressing challenges for developing economies, driven by rapid urbanization, aging infrastructure, and the intensifying impacts of climate change [39, 49]. Traditional centralized drainage systems built on large pipe networks and treatment facilities are often inadequate for managing the variability and frequency of extreme rainfall events [31]. As a result, cities face not only physical damage and economic disruption but also heightened public health and social vulnerabilities. Addressing these challenges requires rethinking drainage design through the lens of resilience, adaptation, and sustainability. Resilient urban drainage design emphasizes the ability of systems to withstand shocks, adapt to unforeseen circumstances, and provide additional benefits beyond flood control [28]. In this context, two strategies are particularly noteworthy: nature-based solutions (NbS) and decentralized systems [17, 35]. Permeable pavements, green roofs, and small retention basins are examples of decentralized strategies that reduce runoff at its source, ease strain on existing networks, and improve flexibility in rapidly growing metropolitan areas [62]. They also provide redundancy by spreading water management tasks across small-scale system, reducing the likelihood of systemic failure in cities. In contrast, NbS use natural processes to restore the hydrological balance and provide additional advantages. Built wetlands, bioswales, and urban green corridors are a few examples that not only reduce peak flows but also

M. S. Islam · B. Sun · J.-Q. Lv · Y. Liu (✉)
State Key Laboratory of Regional Environment and Sustainability, School of Environment, Tsinghua University, Beijing 100084, China
e-mail: liuyc@mail.tsinghua.edu.cn

A. Cheshmehzangi et al. (eds.), *Nature-based and Climate Responsive Design for Urban Systems*, Urban Sustainability, https://doi.org/10.1007/978-981-92-2070-0_4

enhance water quality, promote biodiversity, and offer recreational and social benefits. NbSs are affordable and flexible substitutes for entirely constructed infrastructure in developing nations with limited resources. However, there are still obstacles to adoption, such as limited funding, poor governance, and disjointed institutional frameworks [42]. Active community involvement, more robust policy support, and integrated planning are necessary to overcome these obstacles. This chapter examines how nature-based and decentralized methods might improve urban drainage resilience, opening the door to sustainable urban futures and climate adaptation in developing countries. Both green and gray infrastructures could improve urban water system resilience [19]. Climate change adaptation has become one of the keys to preparing for changes [13]. Urban drainage systems are indeed significantly impacted by current climate change and future trends, which pose both immediate and long-term challenges [1, 30].

This study aims to explore existing research and design in urban drainage and, ultimately, provide insights into future drainage designs in a developing economy.

2 Methodology

This study employs a comprehensive approach that combines a literature review, infrastructure assessment, and case evaluation to explore climate-resilient urban drainage systems in developing megacities. A detailed literature review identified challenges associated with traditional centralized drainage systems, public health impacts, and climate vulnerabilities. Existing evidence from Dhaka was examined to assess sewer coverage deficits and wastewater management challenges. A decentralized and nature-based drainage framework was then conceptualized, incorporating green infrastructure such as green roofs, permeable pavements, community sanitation systems, and constructed wetlands. Selected case studies were analyzed to illustrate the potential effectiveness of these approaches for flood mitigation, water quality improvement, and greenhouse gas reduction. Finally, governance, financing, and community engagement challenges reported in the literature were reviewed, and policy recommendations were proposed to support the implementation of sustainable urban drainage solutions.

3 Challenges of Traditional Drainage in Developing Countries

Rapid economic development has placed pressure on city dwellers regarding waste management, especially urban wastewater management. At present, sewerage services cover approximately 70–95% of the urban population in high-income countries, whereas coverage in low- and middle-income countries remains limited to roughly 10–40% [52]. This stark disparity underscores persistent global inequalities in urban sanitation infrastructure and access to basic public services (Fig. 1). Urban drainage in developing countries faces greater challenges due to outdated sanitation philosophies, space limitations from informal settlements, and severe stormwater pollution, ultimately resulting in severe water pollution in urban water bodies. Additionally, socioeconomic, climatic, and institutional barriers hinder the adoption of modern, sustainable drainage solutions [38, 45]. In Dhaka city, only 20% of the area has sewer lines [57], and only 2% of the total population of Jakarta city is connected to the sewer network [7], while 30% of the total sewerage remains untreated in Mumbai city [46]. In many developing countries, including Bangladesh, Pakistan, and India, informal settlements such as slums lack sewer lines or hygienically maintained sewer systems [12]. In comparison, approximately 80% people in the USA are connected to public sewer lines [58].

In developing and low-income megacities, socioeconomic status is strongly correlated with access to sewerage infrastructure. Slums and economically disadvantaged communities frequently lack connections to centralized sewer systems, resulting in disparities in sanitation access [4]. Consequently, only wealthier households maintain consistent access to formal sewer networks, exacerbating public health and environmental inequities. Figure 1 shows Open drainage carrying domestic wastewater in Indonesia [18] and low-income communities connected with open drainage in Dhaka, Bangladesh.

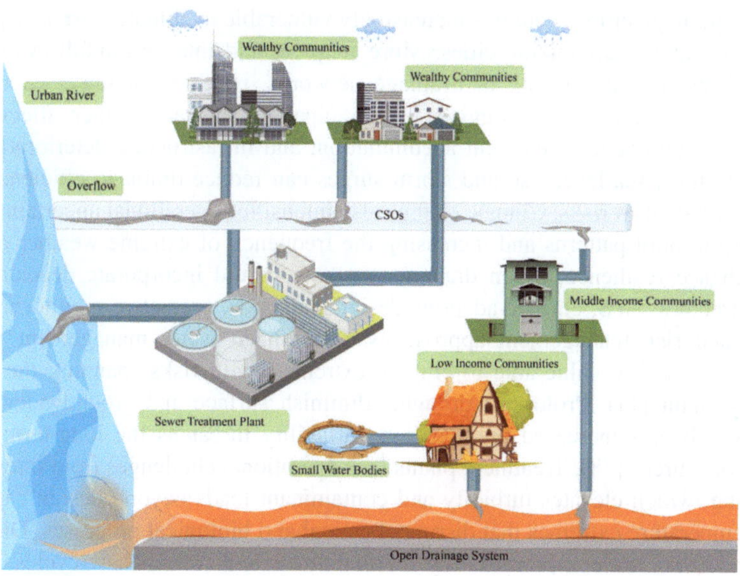

Fig. 1 Schematic representation of the drainage network facilities in Dhaka and Jakarta across different communities based on income

3.1 Centralized Systems and Their Limitations

A centralized drainage system is one of the traditional methods of urban stormwater management [60]. Centralized urban drainage systems include stormwater inlets, large pipe networks, pumping stations, detention and storage facilities, and final discharge outlets. The network must be hydraulically reliable to cope with short-term fluctuations in rainfall and runoff. Pumping and storage facilities regulate flow and reduce peak loads, thereby influencing the hydraulic characteristics of the overall system. Municipal discharge standards define key parameters and limits through regulations. The drainage system must handle the dynamic variability of stormwater flows in both time and space. Still, its performance is subject to disruptions caused by extreme rainfall events, rapid urban runoff from impervious surfaces, pollution accumulation, blockages, pipe leakage, and aging infrastructure. Performance degradation from long-term service further exacerbates operation costs and reliability challenges. The rate of degradation depends on operational and maintenance practices, material properties, environmental stresses, and the growing impact of climate change.

3.2 Vulnerabilities to Climate Extremes

Urban drainage infrastructure is increasingly vulnerable to climate extremes, particularly in rapidly urbanizing cities. More frequent and intense rainfall events can exceed the design capacity of drainage networks, resulting in sewer overflows, urban flooding, and waterlogging. Aging pipe systems are further affected by prolonged inundation, sediment accumulation, and infrastructure deterioration. In coastal cities, sea-level rise and storm surges can reduce drainage efficiency and increase backflow risks. Climate change also intensifies operational uncertainties by altering rainfall patterns and increasing the frequency of extreme weather events. To enhance resilience, urban drainage systems should incorporate decentralized infrastructure, NbS, climate-adaptive design standards, predictive monitoring, and integrated risk management approaches. Centralized water management infrastructure is highly vulnerable to climate extremes, with risks spanning the entire supply chain [11]. Prolonged droughts diminish surface and groundwater availability, whereas increased precipitation variability threatens the long-term reliability of sources [48]. Treatment plants face operational challenges from intensified flooding, which elevates turbidity and contaminant loads beyond design capacity, and temperature extremes that disrupt chemical and biological processes. Distribution networks are compromised by thermal stress and freeze–thaw cycles, which accelerate pipe degradation, as well as by inundation from extreme rainfall, which can cause hydraulic failure. Energy-dependent pumping systems are susceptible to outages during storms or heatwaves, exacerbating supply–demand imbalances during compound events. Long-term climate stressors, including saltwater intrusion from sea-level rise and permafrost thaw-induced infrastructure instability in northern regions, further threaten system integrity. To increase resilience, decentralized backup systems, climate-adaptive materials, predictive monitoring, and integrated risk assessments are needed to address these multidimensional challenges under evolving climatic conditions.

3.3 Infrastructure Underfunding and Urban Sprawl

Urban sprawl and inadequate investment in drainage infrastructure pose significant problems for public health and municipal management in developing nations [21, 57]. Rapid population growth in many places puts tremendous strain on antiquated drainage systems that receive little funding or upkeep. Because of this, cities often struggle to handle excessive rainfall, which usually leads to flooding and waterlogging. Public health is at risk due to poor drainage infrastructure, which contributes to water contamination and the spread of waterborne illnesses. Urban sprawl, driven by the unplanned and rapid expansion of cities into surrounding rural areas, exacerbates the problem. More surface runoff results from the replacement of more natural land with concrete structures like buildings and roads, which reduces the ground's capacity to absorb rainfall. Effective stormwater management is hampered by this, which not only overburdens current drainage systems but also results in disjointed, ineffective infrastructure networks. Unchecked urban growth and underfunded drainage systems combine to create a vicious cycle in which poor infrastructure exacerbates environmental degradation and heightens vulnerability to climate change. To address these problems, developing nations require thorough urban planning that combines investments in resilient drainage systems with sustainable growth, making cities ready to face new challenges and enhancing the standard of living for citizens. Figure 2 indicates significant disparities in water, sanitation, and wastewater service provision, with full coverage achieved only for the water network, while sewerage coverage remains limited to 20%. A large proportion of the population (80%) relies on onsite sanitation systems; however, only 54% have access to improved containment, and nearly half depend on shared facilities. The treatment performance is remarkably low, with only 9.6% wastewater treated and no recorded sludge treatment, indicating critical gaps in post-collection management.

The high dependence on non-networked sanitation, coupled with limited treatment capacity, suggests substantial risks to environmental quality and public health, as untreated waste streams likely contribute to pathogen dissemination and pollutant loading in receiving environments. Addressing these gaps will require both infrastructure expansion and improvements in management.

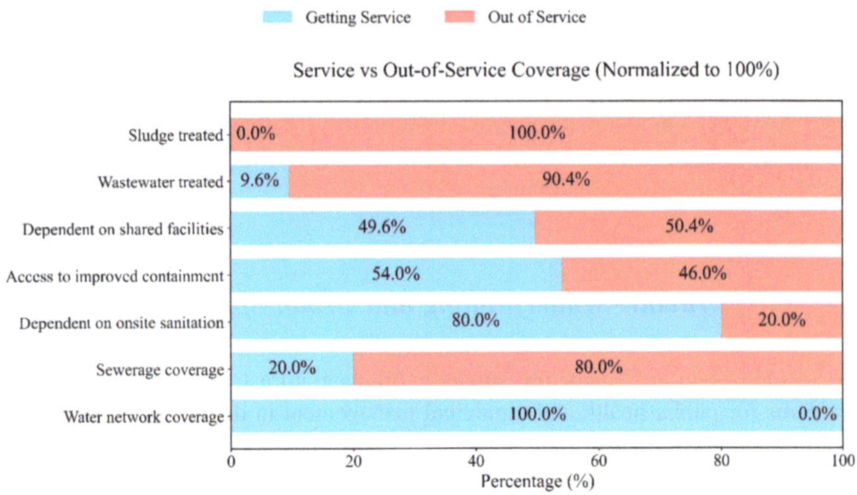

Fig. 2 Wastewater services coverage of Dhaka city (Data were adopted from [41])

3.4 Public Health and Pollution Impacts

Public health and environmental pollution are closely linked to the quality of sanitation and drainage infrastructure. A lack of adequate sewage treatment and drainage systems can lead to water pollution, disease transmission, and ecological damage. For example, in India, owing to inadequate sanitation, millions of people still practice open defecation, resulting in fecal contamination of groundwater and widespread diarrheal and cholera outbreaks [10]. Similarly, in sub-Saharan Africa, inadequate drainage systems allow stagnant water to become mosquito breeding grounds during the rainy season, significantly increasing the risk of malaria and dengue fever transmission [3, 14]. With respect to environmental pollution, the discharge of untreated sewage leads to eutrophication of water bodies [55]. For example, the discharge of industrial and domestic wastewater in five urban areas severely affects water quality [57]. Furthermore, clogged or poorly designed drainage systems can exacerbate urban flooding. For example, during the 2021 torrential rainstorm in Zhengzhou, insufficient drainage capacity led to severe flooding, causing not only economic losses but also an increased risk of waterborne infectious diseases [54]. Addressing these issues requires a multipronged approach, including investment in wastewater treatment facilities, optimizing urban drainage design such as Singapore's "Active, Beautiful, Clean" ABC Waters program [56], and strengthening regulatory policies (such as the EU Water Framework Directive's strict restrictions on industrial wastewater discharge). Research shows that for every \$1 invested in sanitation, \$4.30 in health care expenses can be saved [43], highlighting the dual economic and health benefits of improving infrastructure.

3.4.1 Untreated Overflows and Waterborne Disease

The lack of sustainable sewage treatment infrastructure results in frequent untreated sewage overflows, significantly increasing the risk of waterborne diseases. When heavy rains occur or systems are overloaded, mixed sewage is discharged directly into surface water bodies, contaminating drinking water sources and spreading pathogens such as Vibrio cholerae, Salmonella typhi, and pathogenic *Escherichia coli*. Epidemiological studies have shown that in urban areas with aging drainage systems, the incidence of acute watery diarrhea during the rainy season is typically 40–60% higher than that during the dry season. This risk is particularly acute in low-income communities, which often lack adequate water supply networks and rely more on potentially contaminated surface water or shallow groundwater. Furthermore, chronic stagnant water provides ideal breeding grounds for disease vectors such as the malaria mosquito and the dengue vector, Aedes, exacerbating the disease burden. From an environmental health perspective, persistent sewage overflows can also lead to eutrophication, damage aquatic ecosystems, and potentially trigger broader public health issues through the food chain, such as contaminated seafood. To effectively block this transmission chain, it is necessary to establish a climate-adaptive hierarchical drainage system and combine it with NbS such as artificial wetlands for end-of-pipe treatment, thereby improving the health and safety of the water environment at the system level.

3.4.2 Greenhouse Gas Emissions from Anaerobic Sewer Zones

GHG emission sources in urban water systems include septic tanks, sewer pipe networks, open channels, and artificial wetlands [15]. These infrastructures produce large amounts of potent greenhouse gases, such as CH_4 and N_2O, through microbial metabolism during the degradation of organic matter (Fig. 3). Among them, septic tanks exhibit high CH_4 and CO_2 yields due to the anaerobic environment [44]. Pressure-sewage pipe networks promote the release of dissolved GHGs under turbulent conditions; open-channel systems provide important pathways for GHG emissions through air–water interface exchange. Though artificial wetlands have carbon sink functions, their anaerobic–aerobic alternating environment significantly promotes the generation of N_2O, and the emission flux can reach 25–120 mg $N_2O/m^2 \cdot h$.

These emission processes are regulated by multiple environmental factors, such as water temperature, hydraulic retention time, redox potential, and the carbon–nitrogen ratio, and need to be accurately quantified through a combination of on-site monitoring and model simulation.

Fig. 3 Greenhouse gas emissions from urban drainage and associated infrastructures

4 Design Principles and Enabling Approaches: Decentralized and NbS

Decentralized water management is the process of distributing water governance, infrastructure, and services to local or regional levels to improve efficiency, responsiveness, resilience, and sustainability in addressing water-related challenges [24] (Fig. 4).

The diagram summarizes the core elements of modern urban drainage infrastructure for water management, including overflow drainage systems to prevent flooding, separate collection and treatment of blackwater and graywater to improve wastewater treatment efficiency, rainwater infiltration and reuse to promote natural water circulation and support agricultural irrigation, and a management model that combines centralized traditional pipe network systems with decentralized green infrastructure such as rain gardens. This system not only ensures urban flood control but also optimizes water resource utilization, embodying the concepts of SuDSs and LID, with the aim of building a resilient urban water environment.

4.1 Green Infrastructure

Urban green drainage infrastructure is a sustainable stormwater management solution that enhances urban resilience and ecosystem services through nature-based solutions [36]. This decentralized system uses permeable paving, bioretention ponds,

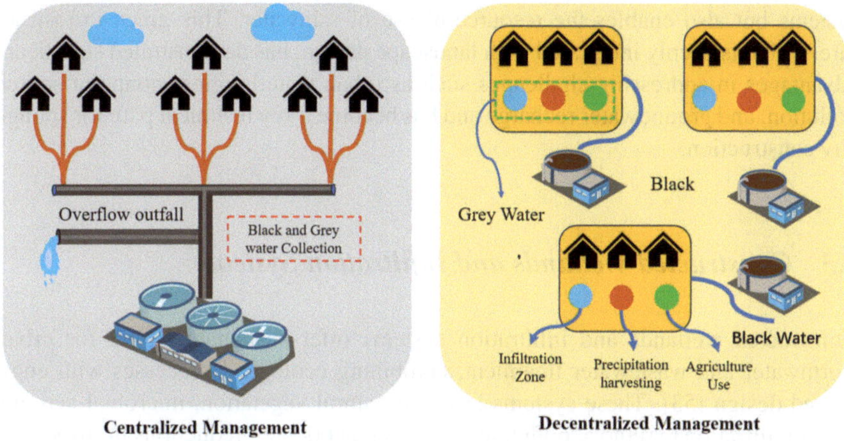

Fig. 4 Concept of a centralized and decentralized water management system and infrastructure

rain gardens, constructed wetlands, and green roofs to mimic natural hydrological processes, reducing runoff and peak flows while improving water quality through filtration and bioremediation. The replacement of traditional gray infrastructure with vegetation systems effectively mitigates urban flood risk, promotes groundwater recharge, and reduces the burden on drainage networks. This multifunctional design also offers co-benefits, including mitigating the urban heat island effect, sequestering carbon and releasing oxygen, creating habitats, and increasing landscape value. From rooftop retention ponds to park-scale storage tanks, the strategic deployment of green drainage infrastructure across the watershed has formed a "sponge city" network that adapts to climate change. Combined with intelligent monitoring and adaptive governance, green drainage infrastructure demonstrates how ecological engineering can be transformed into opportunities for urban water governance, driving cities toward regenerative development.

4.2 Low-Impact Development (LID)

Low-impact development (LID) is an urban design strategy that controls stormwater runoff at its source [20]. Its core concept is to simulate the natural hydrological cycle through decentralized, small-scale ecological measures. This technical system incorporates typical measures such as bioretention facilities, permeable pavement, and vegetated shallow trenches, achieving site-scale stormwater retention, infiltration, and purification, effectively reducing runoff peaks and pollution loads. Unlike traditional centralized end-of-pipe treatment models, LID emphasizes maintaining or restoring the site's natural hydrological functions during the development process. By constructing a multilevel control system comprising "source reduction, process control, and end-of-pipe regulation," it not only alleviates pressure on urban drainage systems but also enables the resourceful use of rainwater. This green infrastructure, which is deeply integrated with landscape design, has demonstrated significant advantages in addressing challenges such as urban waterlogging, nonpoint source pollution, and groundwater recharge and has become a key technical path for sponge city construction.

4.3 Constructed Wetlands and Infiltration Systems

Constructed wetlands and infiltration systems offer sustainable, NbS for urban stormwater and wastewater treatment, combining ecological processes with engineered design [53]. These systems leverage natural vegetation, microbial activity, and soil filtration to remove pollutants, resulting in 60–90% reductions in suspended solids, nitrogen, and pathogens. Case studies, such as Kolkata's East Wetlands (treating 750 MLDs of sewage) and Singapore's rain gardens (reducing peak flows by 30%), have demonstrated their scalability [32]. Operational principles include controlled hydraulic loading, substrate selection (e.g., gravel, biochar), and phased anaerobic/aerobic zones to optimize treatment. Compared with conventional systems, they require 40% lower operational costs while enhancing biodiversity and groundwater recharge, making them particularly viable for decentralized applications in developing cities. In conclusion, the design principles that drive decentralization and NbS have been explained in this section, along with how these strategies work as an enabling toolbox to achieve key performance goals (runoff reduction, water quality improvement, resilience, and co-benefits). However, a city's operational drainage system configuration is not determined solely by principles. Therefore, the next stage is to convert these ideas into an organized design framework that connects goals to component selection, spatial arrangement, and capacity distribution within practical constraints (land availability, informality, maintenance capacity, and climate variability). Consequently, an integrated SuDS design logic that links principle-level decisions to system-level layout and sizing decisions is presented in Sect. 5.

5 SuDS Design Framework: Translating Principles into an Integrated Gray-Green System

This section creates a system-level SuDS design framework for urban drainage in developing-city contexts, building directly on the enabling strategies and principles presented in Sect. 4. We arrange the design as a multilayer, integrated gray-green system which includes surface, subsurface, and ecological layers and demonstrate how certain principles (source control, decentralization, retention/detention, treatment, and safe conveyance) are operationalized at each layer. In practice, this structure offers a clear route from "what the system should achieve" to "how it should be configured." SuDS focuses on managing stormwater sustainably at a local scale through nature-based solutions such as green roofs and permeable pavements to mimic natural drainage and reduce flooding [37].

5.1 Framework for Multilayered Drainage

Figure 5 shows the schematic diagram illustrating a multifunctional SuDS that integrates surface, subsurface, and ecological layers. It aims to improve stormwater management in densely populated areas and mitigate the risk of urban flooding.

At the surface level, green roofs on residential buildings reduce stormwater runoff, whereas permeable paving promotes rainwater infiltration and reduces surface water accumulation. Silt traps and sediment forebays are located at the front end of the system to intercept coarse particulate pollutants and ensure the efficiency of subsequent treatment units. The subsurface layer includes a rainwater retention tank connected to the combined sewer system. A smart sensor flow control system monitors and regulates discharge flow in real time, optimizing pipe network capacity utilization and effectively preventing overflow. Furthermore, rooftop rainwater harvesting and reuse devices enable water recycling and improve water resource efficiency. At the ecological level, tree trenches and bioswales are installed to increase rainwater purification and groundwater recharge through natural filtration and nutrient absorption by plant roots. Constructed wetlands are located at the system's terminal, using phytoremediation to purify tailwater and provide habitat for

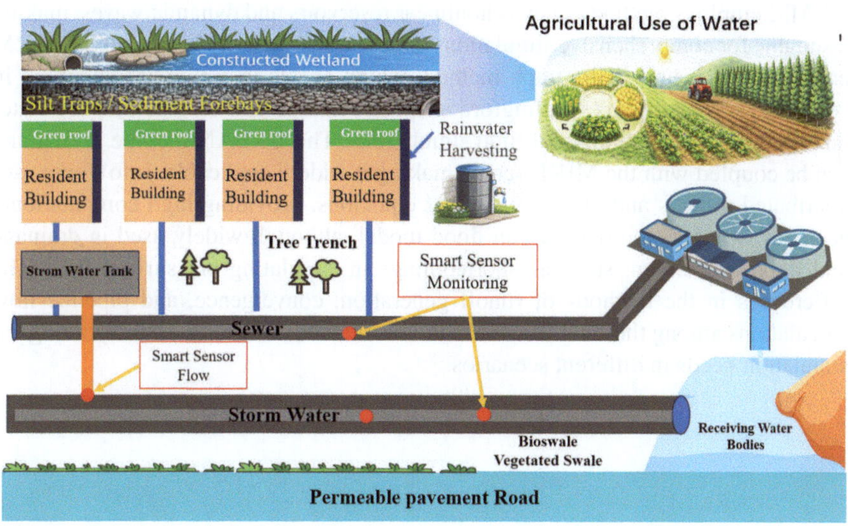

Fig. 5 Schematic diagram of the integrated gray-green integration of surface, subsurface, and biological layers of urban infrastructure

urban biodiversity. This integrated design, through a "gray-green" approach, achieves synergistic optimization of drainage and flood prevention with ecological restoration, meeting the practical needs of sustainable water management in developing cities.

5.2 Modeling and Simulation Tools

In the design, control, and management of urban drainage systems, modeling and simulation are crucial, particularly for data generation [23]. The main reasons we need urban drainage models include the following: first, models can simulate the system's response under different rainfall scenarios, helping to predict flooding risk points; second, simulations can evaluate the capacity limitations and bottlenecks of existing drainage systems; moreover, models provide a scientific basis for planning new drainage infrastructure or renovating existing systems; additionally, in the context of climate change leading to frequent extreme weather, models can test the adaptability and resilience of the system; finally, simulated data can support the development of intelligent drainage systems and the formulation of optimized operational strategies. These functions make models indispensable tools for the scientific management and decision support of urban drainage systems, especially in the absence of long-term measured data, where model simulation becomes an important means of obtaining system performance data (Table 1).

Modeling and monitoring to understand rainfall-runoff patterns help SuDS design [16, 33]. Urban flood models such as SWMM, STORM, Wallingford, URBAN, InfoWorks, and the plain urban flood model each have unique features (Table 1). SWMM employs methods such as nonlinear reservoirs and dynamic waves, making it suitable for comprehensive simulations of both water quantity and quality. STORM is based on the unit hydrograph method, which excels in simulating overflow in combined sewer systems. Wallingford integrates rainfall-runoff, pipeline, and water quality modules, enabling real-time simulations. The URBAN module is rich and can be coupled with the MIKE series, making it widely applicable. InfoWorks uses distributed models and the Saint–Venant equations, providing high computational accuracy, whereas the plain urban flood model, although widely used in drainage and flood prevention, still has shortcomings in simulating pressurized flow. The differences in the methods of runoff generation, convergence, and pipeline flow calculations among these models make them suitable for various urban hydrological simulation needs in different scenarios.

Table 1 Modeling of urban water infrastructures and their functions

Model name	Runoff module	Confluence module	Pipeline module	Key features	References
SWMM	Infiltration curve method, SCS method	Nonlinear storage	Steady/unsteady flow, kinematic/dynamic wave	Dynamic rainfall-runoff modeling is suitable for water quality and is widely used	[34, 50]
STORM	SCS method, rainfall loss	Unit hydrograph	Hydrological routing methods	Urban flood and quality simulation; suitable for combined/separated systems	[25]
Wallingford	Nonlinear reservoir, storage functions	Muskingum and hidden routing	SWMM equivalent hydraulic modules	Macro simulation supports water quality and pipe design	[9]
URBAN	Rainfall infiltration	Kinematic/dynamic wave	Dynamic pipe and real-time control	Integrated pipe design, control, and real-time simulation	[8]
InfoWorks	Fixed ratio method, SCS, regression curves	Double reservoir + SWMM	Saint–Venant equations	Short-term simulation with rainfall-runoff and pipe hydraulics	[51]
Visualized Data Generator	I&I via instantaneous unit hydrograph + conductivity	Multi-level tree (residential, industrial, PS, WWTP)	Delay + Fourier transport	Visualized, pollutant-hydrograph, Fourier variation, real-data calibrated	[33]
ML Model	Data-driven (precipitation, evapotranspiration, temperature, land cover inputs)	Implicit nonlinear mapping	Implicit in hidden layers	Machine-learning based; self-attention improves interpretability and performance; less dependent on	[29]

6 Discussion and Implications

6.1 Construction of the Sewage Pipeline

The population density of a city varies from place to place. Therefore, before constructing a depiction of the population distribution, it is imperative to measure the length of the sewage line necessary. The failure rate of the sewage pipe network can be a potential measurement option [54]. Generally, the length of the sewage pipeline should be equal to or greater than the water supply. These two systems have a symmetric relationship. According to the annual report of Dhaka WASA, the length of the water supply line is 3875 km, with a capacity of 2650–2700 million liters of water every day, whereas the sewer line is only 934 km long. Using sewer-to-water network length as a proxy indicator, the sewer coverage deficit is approximately 75.9%. This clearly reflects that a significant amount of wastewater was transported at the surface and ultimately left the urban water bodies untreated. The incorporation of a sewer line with a water supply line is imperative. In addition to the water supply facility area, more than 4.4 million people live in the slums of the capital. These slums can be considered poor and low-income urban settlements in Dhaka city. Building community-based septic tanks and connecting them to the main sewer network will be a cost-effective solution for this type of low-income settlement, as many of these settlements are informal and lack access to formal sanitation infrastructure [26].

6.2 Infrastructure Standards

The standards of sewer infrastructure can vary among different countries, depending on factors such as the local geography, climate, population density, and regulatory environment. In general, developed countries tend to have more advanced sewer infrastructure than developing countries do, due to higher levels of investment, technological innovation, and effective governance. In China, the standards of sewer infrastructure have improved significantly over the past few decades due to increased investment in water and wastewater treatment facilities and stronger regulatory enforcement. According to the National Urban Wastewater Treatment and Recycling Capacity Promotion Plan, China aims to achieve 95% coverage of urban wastewater treatment by 2030 and to increase the proportion of recycled water to 25% of total urban water use. In developed countries such as the United States, sewer infrastructure is typically designed to meet high performance and environmental protection standards. For example, the Environmental Protection Agency (EPA) has set national standards for the discharge of pollutants into waterways and requires municipal wastewater treatment plants to meet specific design and performance criteria. In addition, many cities in the United States have implemented innovative sewer infrastructure, such as green roofs, rain gardens, and permeable pavement, to reduce stormwater runoff and improve water quality.

6.3 WRRFs System Development

Water resource recovery facility (WRRFs) are dedicated systems designed to treat and reuse water for non-potable applications, including industrial processes, irrigation, and toilet flushing. These facilities are instrumental in alleviating stress on freshwater reserves and enhancing the efficiency of treatment processes. In the context of Dhaka, where mounting strains on freshwater reserves are evident, WRRFs play a pivotal role. Both domestic and industrial wastewater severely pollute rivers encircling cities, while excessive groundwater extraction for drinking and domestic use has led to groundwater depletion. To develop WRRFs, it is important to construct satellite treatment facilities so that wastewater can be used locally immediately after treatment.

6.4 Remove Illegal Connections

Illegal sewer connections are common in the megacities of developing countries [55] and cause serious urban water pollution [27]. The scenario has worsened in Dhaka city. Most houses have illegal connections to surface drainage, which is constructed to transport urban stormwater to urban water bodies. However, owing to illegal domestic and industrial wastewater connections, domestic pollutants ultimately reach urban water bodies. Recently, Dhaka North City Corporation reported that 3265 houses out of 3830 discharge their wastewater directly to surface stormwater drainage through illegal connections, whereas 524 houses partially have illegal connections to surface water. Only 41 houses have proper sewer connections to the sanitary sewer system. Even in the weather, some cities have similar situations. To protect urban health and ecology, it is imperative to identify illegal connections and correct them.

6.5 Nature-Based Solutions

Nature-based solutions (NbS) involve the sustainable management of natural elements and features to solve socioenvironmental problems. The NBS method could be a potential source for increasing environmental services and addressing the problems associated with urban water infrastructure, especially wastewater and stormwater. Extreme climate events, especially precipitation events, pressure the sewer network [2, 6]. Rooftop farming is becoming popular in Dhaka, Bangladesh. Mastura [40] estimated that, for two households in Dhaka city, stormwater management benefits from rooftop farming by the practitioner in Mirpur, the Dhaka area, range between US$ 361 and US$ 11,494 [40]. Although rooftop farming can play an important role in mitigating urban stormwater flooding, several challenges remain, including space shortages, a lack of technical knowledge, and limited leisure time. If the government and other stakeholders emphasize rooftop farming, which might be a potential solution and asset for urban wellbeing.

6.6 GHG Mitigation Strategies

The physical design and structure of the sewer pipes have a significant impact on GHG accumulation and emission levels [59]. GHG emission reduction strategies for urban drainage systems can be achieved through optimized design: a diversion system collects blackwater (with high COD) and graywater (low load) to reduce wastewater treatment energy consumption and methane emissions; promotes green infrastructure such as rain gardens and infiltration trenches to increase carbon sequestration and reduce pumping energy consumption; utilizes anaerobic digestion to recover biogas energy from sludge and deploys nitrogen and phosphorus removal processes at centralized treatment plants to prevent N_2O emissions. Through a hybrid "centralized-decentralized" management model, combined with low-carbon materials and intelligent monitoring systems, drainage systems can reduce carbon emissions throughout their lifecycles and contribute to carbon-neutrality goals.

6.7 Source Control Measures

Vegetated roof surfaces, commonly known as green roofs, can serve as effective source control measures by absorbing and retaining rainwater at the point of precipitation. A recent study found that approximately 77% of rooftop areas in Dhaka city are suitable or highly suitable for rooftop agriculture [47], highlighting the significant potential for integrating green roofs into urban stormwater management strategies. Rooftop agriculture is becoming popular in high-income countries [5]; however, it may be a viable option for megacities in developing countries. Bangladesh receives approximately 2200 mm of precipitation annually, which has become a major driver of urban waterlogging. Pollution source control methods have been studied in recent years. For example, nitrates produced in urine can be controlled to reduce H_2S odors in sewers [61].

7 Conclusion

This chapter emphasizes the importance of sustainable urban drainage systems (SuDS) in boosting climate resilience in emerging cities. Urban areas can significantly reduce greenhouse gas emissions, improve water quality, and mitigate flooding by combining NbS with decentralized drainage systems. It is clear from real-world case studies and computer models that decentralized systems, such as green infrastructure and LID strategies, provide adaptable, affordable solutions, particularly in settings with limited resources. The application of SuDS can result in more robust and adaptable urban ecosystems, notwithstanding obstacles such as insufficient funding, poor governance, and disjointed institutional frameworks. The study emphasizes the importance of integrating ecological engineering and urban design to achieve long-term sustainability and climate adaptation objectives. To sum up, implementing decentralized and nature-based strategies can help to solve the current issues caused by climate change while also opening the door for future low-carbon, more resilient communities.

Acknowledgements This research was supported by the National Natural Science Foundation of China (no. 52425002, 52270091, W2442025).

References

1. Abebe Y, Tesfamariam S (2019) Climate change impact and adaptation for urban drainage systems. In: Bastidas-Arteaga E, Stewar MG (eds) Climate adaptation engineering. Butterworth-Heinemann, Chapter 3, pp 73–98. https://doi.org/10.1016/B978-0-12-816782-3.00003-6
2. Addison-Atkinson W, Chen AS, Memon FA, Chang T-J (2022) Modelling urban sewer flooding and quantitative microbial risk assessment: A critical review. J Flood Risk Manag 15(4):e12844. https://doi.org/10.1111/jfr3.12844
3. Agbo C, Jeffrey P, Sule MN (2025) Evaluation of failings in urban water supply and sanitation systems in Sub-Saharan Africa: a systematic review to inform future planning. J Water, SanitIon Hyg Dev 15(2):148–165. https://doi.org/10.2166/washdev.2025.267
4. Alam M-U, Rahat MA, Neeher N, Tabassum KF, Nawaz S, Evans B, Hutchings P (2025) Barriers to sewer connection in urban Dhaka: individual and organisational challenges. J Water, SanitIon Hyg Dev 15(4):287–304. https://doi.org/10.2166/washdev.2025.281
5. Appolloni E, Orsini F, Specht K, Thomaier S, Sanyé-Mengual E, Pennisi G, Gianquinto G (2021) The global rise of urban rooftop agriculture: A review of worldwide cases. J Clean Prod 296:126556. https://doi.org/10.1016/j.jclepro.2021.126556
6. Arthur PMA, Konaté Y, Sawadogo B, Sagoe G, Dwumfour-Asare B, Ahmed I, Williams MNV (2022) Performance evaluation of a full-scale upflow anaerobic sludge blanket reactor coupled with trickling filters for municipal wastewater treatment in a developing country. Heliyon 8(8):e10129. https://doi.org/10.1016/j.heliyon.2022.e10129
7. Bakhshipour Amin E, Dittmer U, Haghighi A, Nowak W (2021) Toward sustainable urban drainage infrastructure planning: a combined multiobjective optimization and multicriteria decision-making platform. J Water Resour Plan Manag 147(8):04021049. https://doi.org/10.1061/(ASCE)WR.1943-5452.0001389
8. Balla KM, Schou C, Bendtsen JD, Ocampo-Martinez C, Kallesøe CS (2022) A nonlinear predictive control approach for urban drainage networks using data-driven models and moving horizon estimation. IEEE Trans Control Syst Technol 30(5):2147–2162. https://doi.org/10.1109/TCST.2021.3137712
9. Beran M (1987) Institute of hydrology wallingford. Flood Hydrology: Proceeding of the international symposium on flood frequency and risk analyses, 14–17 May 1986, Louisiana State University, Baton Rouge, USA
10. Biswas S, Adhikary M, Alam A, Islam N, Roy R (2024) Disparities in access to water, sanitation, and hygiene (WASH) services and the status of SDG-6 implementation across districts and states in India. Heliyon 10(18):e37646. https://doi.org/10.1016/j.heliyon.2024.e37646
11. Borzì I (2023) Vulnerability assessment of water supply infrastructures through multiple indicator methodology. J Water Clim Chang 14(11):3967–3984. https://doi.org/10.2166/wcc.2023.148

12. Brouwer R, Sharmin DF, Elliott S, Liu J, Khan MR (2023) Costs and benefits of improving water and sanitation in slums and non-slum neighborhoods in Dhaka, a fast-growing mega-city. Ecol Econ 207:107763. https://doi.org/10.1016/j.ecolecon.2023.107763
13. Callahan CW (2025) Present and future limits to climate change adaptation. Nat Sustain 8(4):336–342. https://doi.org/10.1038/s41893-025-01519-7
14. Chan K, Tusting LS, Bottomley C, Saito K, Djouaka R, Lines J (2022) Malaria transmission and prevalence in rice-growing versus non-rice-growing villages in Africa: a systematic review and meta-analysis. Lancet Planet Health 6(3):e257–e269. https://doi.org/10.1016/S2542-5196(21)00349-1
15. Chen J, Wang H, Yin W, Wang Y, Lv J, Wang A (2024) Deciphering carbon emissions in urban sewer networks: Bridging urban sewer networks with city-wide environmental dynamics. Water Res 256:121576. https://doi.org/10.1016/j.watres.2024.121576
16. Chen SS, Tsang DCW, He M, Sun Y, Lau LSY, Leung RWM, Lau ESC, Hou D, Liu A, Mohanty S (2021) Designing sustainable drainage systems in subtropical cities: Challenges and opportunities. J Clean Prod 280:124418. https://doi.org/10.1016/j.jclepro.2020.124418
17. Debele SE, Leo LS, Kumar P, Sahani J, Ommer J, Bucchignani E, Vranić S, Kalas M, Amirzada Z, Pavlova I, Shah MAR, Gonzalez-Ollauri A, Di Sabatino S (2023) Nature-based solutions can help reduce the impact of natural hazards: A global analysis of NBS case studies. Sci Total Environ 902:165824. https://doi.org/10.1016/j.scitotenv.2023.165824
18. Desk N (2023) Jakarta commences sewer construction. Jakarta Post
19. Dong X, Guo H, Zeng S (2017) Enhancing future resilience in urban drainage system: Green versus grey infrastructure. Water Res 124:280–289. https://doi.org/10.1016/j.watres.2017.07.038
20. Essamlali I, Nhaila H, El Khaili M (2024) A new architecture of Low Impact Development (LID)-based stormwater management system through Internet of Things (IoT) and Machine Learning integration. Case Stud Chem Environ Eng 10:100942. https://doi.org/10.1016/j.cscee.2024.100942
21. Ferdowsi A, Behzadian K (2024) Urban drainage infrastructures toward a sustainable future. In: Bahrami A (ed) Sustainable structures and buildings. Springer International Publishing, pp 111–119. https://doi.org/10.1007/978-3-031-46688-5_8
22. Francisco THS, Menezes OVC, Guedes ALA, Maquera G, Neto DCV, Longo OC, Chinelli CK, Soares CAP (2023) The main challenges for improving urban drainage systems from the perspective of brazilian professionals. Infrastructures 8(1)
23. García L, Barreiro-Gomez J, Escobar E, Téllez D, Quijano N, Ocampo-Martinez C (2015) Modeling and real-time control of urban drainage systems: A review. Adv Water Resour 85:120–132. https://doi.org/10.1016/j.advwatres.2015.08.007
24. Garrido-Baserba M, Sedlak DL, Molinos-Senante M, Barnosell I, Schraa O, Rosso D, Verdaguer M, Poch M (2024) Using water and wastewater decentralization to enhance the resilience and sustainability of cities. Nat Water 2(10):953–974. https://doi.org/10.1038/s44221-024-00303-9
25. Guo Y, Park D, Mejia A, Gironás J, Roesner L (2019). Probabilistic models for urban stormwater management. In: Statistical analysis of hydrologic variables, pp 333–380. https://doi.org/10.1061/9780784415177.ch09
26. Haque AN (2021) Climate risk responses and the urban poor in the global South: the case of Dhaka's flood risk in the low-income settlements. Int J Disaster Risk Reduct 64:102534. https://doi.org/10.1016/j.ijdrr.2021.102534
27. Hoes OAC, Schilperoort RPS, Luxemburg WMJ, Clemens FHLR, van de Giesen NC (2009) Locating illicit connections in storm water sewers using fiber-optic distributed temperature sensing. Water Res 43(20):5187–5197. https://doi.org/10.1016/j.watres.2009.08.020
28. Hosseini S, Barker K, Ramirez-Marquez JE (2016) A review of definitions and measures of system resilience. Reliab Eng & Syst Saf 145:47–61. https://doi.org/10.1016/j.ress.2015.08.006
29. Hu N, Shu K, Zhang Y, Alfonso L, Li C, Zheng T (2025) Prediction of natural runoff in China based on multi-scenario climate models with self-attention neural networks. Water Res 282:123768. https://doi.org/10.1016/j.watres.2025.123768

30. Kourtis IM, Tsihrintzis VA (2021) Adaptation of urban drainage networks to climate change: A review. Sci Total Environ 771:145431. https://doi.org/10.1016/j.scitotenv.2021.145431
31. Lebu S, Gyimah R, Nandoya E, Brown J, Salzberg A, Manga M (2024) Assessment of sanitation infrastructure resilience to extreme rainfall and flooding: Evidence from an informal settlement in Kenya. J Environ Manag 354:120264. https://doi.org/10.1016/j.jenvman.2024.120264
32. Lim HS, Lu XX (2016) Sustainable urban stormwater management in the tropics: An evaluation of Singapore's ABC Waters Program. J Hydrol 538:842–862. https://doi.org/10.1016/j.jhydrol.2016.04.063
33. Liu Y, Islam MS, Chen J, Zhang M, Min H, Li B, Mao J, Tang D, Wang M, Shi H, Huang X (2024) Visualized generator for the simulation of wastewater quality and quantity variations in the sewer system. Sci Total Environ 955:176719. https://doi.org/10.1016/j.scitotenv.2024.176719
34. Martínez C, Vojinovic Z, Price R, Sanchez A (2021) Modelling Infiltration Process, Overland Flow and Sewer System Interactions for Urban Flood Mitigation. Water 13(15):2028
35. Masi F, Langergraber G, Santoni M, Istenič D, Atanasova N, Buttiglieri G (2020) Possibilities of nature-based and hybrid decentralized solutions for reclaimed water reuse. In: Verlicchi P (ed) Advances in chemical pollution, environmental management and protection, vol 5. Elsevier, pp 145–187. https://doi.org/10.1016/bs.apmp.2020.07.004
36. Mitra S, Madhuvanthi S, Sabumon PC (2025) Nature-Based Urban Resilience: Integrating Green Infrastructure. In: Singh P, Srivastava P, Sorokin A (eds) Nature-Based Solutions in Achieving Sustainable Development Goals: Harmonizing Nature and Progress. Springer Nature Switzerland, pp 167–205. https://doi.org/10.1007/978-3-031-76128-7_6
37. Ortega AD, Rodríguez JP, Bharati L (2023) Building flood-resilient cities by promoting SUDS adoption: a multi-sector analysis of barriers and benefits in Bogotá, Colombia. Int J Disaster Risk Reduct 88:103621. https://doi.org/10.1016/j.ijdrr.2023.103621
38. Parkinson J (2002) Urban drainage in developing countries-challenges and opportunities. WATERLINES-LONDON 20(4):2–5
39. Parmesan C, Morecroft MD, Trisurat Y (2022). Climate change 2022: impacts, adaptation and vulnerability
40. Safayet M, Arefin MF, Hasan MMU (2017) Present practice and future prospect of rooftop farming in Dhaka city: a step towards urban sustainability. J Urban Manag 6(2):56–65. https://doi.org/10.1016/j.jum.2017.12.001
41. Saha P, Drabble S, Hanoum R (2023) Evolution of the Dhaka sanitationsector: institutional and regulatoryreform. IWA.
42. Saikia P, Beane G, Garriga RG, Avello P, Ellis L, Fisher S, Leten J, Ruiz-Apilánez I, Shouler M, Ward R, Jiménez A (2022) City Water Resilience Framework: A governance based planning tool to enhance urban water resilience. Sustain Cities Soc 77:103497. https://doi.org/10.1016/j.scs.2021.103497
43. Schermbrucker K (2014) Every dollar invested in water, sanitation brings four-fold return in costs–UN. UN News [New York], 19
44. Shukuru BN, Politaeva NA (2025) Mitigating CH4 and N2O emissions from domestic and industrial wastewater. Renew Sustain Energy Rev 210:115203. https://doi.org/10.1016/j.rser.2024.115203
45. Silveira ALL (2002) Problems of modern urban drainage in developing countries. Water Sci Technol 45(7):31–40. https://doi.org/10.2166/wst.2002.0114
46. Singh A, Kazmi A, Starkl M, Sayanekar S, Herlekar M (2018) Sewage management challenges in mega cities in India: a case study of Mumbai. Desalination Water Treat 116:329–341. https://doi.org/10.5004/dwt.2018.22448
47. Sultana N, Sharifi A, Haque MN, Aghaloo K (2024) Urban greening in Dhaka: Assessing rooftop agriculture suitability using GIS and MCDM techniques. J Environ Manag 368:122146. https://doi.org/10.1016/j.jenvman.2024.122146
48. Tang T, Ge J, Shi H, Wang L, Cao J, Lee X (2025) Drought frequency, intensity, and exposure have increased due to historical land use and land cover changes. Commun Earth & Environ 6(1):398. https://doi.org/10.1038/s43247-025-02392-0

49. Tripathy SS, Chaudhuri S, Murtugudde R, Mhatre V, Parmar D, Pinto M, Zope PE, Dixit V, Karmakar S, Ghosh S (2024) Analysis of Mumbai floods in recent years with crowdsourced data. Urban Clim 53:101815. https://doi.org/10.1016/j.uclim.2024.101815
50. Tu M-C, Wadzuk B, Traver R (2020) Methodology to simulate unsaturated zone hydrology in Storm Water Management Model (SWMM) for green infrastructure design and evaluation. PLoS ONE 15(7):e0235528. https://doi.org/10.1371/journal.pone.0235528
51. Wang K, Chen J, Hu H, Tang Y, Huang J, Wu Y, Lu J, Zhou J (2024) Urban waterlogging simulation and disaster risk analysis using infoworks integrated catchment management: a case study from the Yushan Lake area of Ma'anshan City in China. Water 16(23):3383
52. World Health Organization (2023) Progress on household drinking water, sanitation and hygiene 2000–2022: special focus on gender
53. Wu H, Wang R, Yan P, Wu S, Chen Z, Zhao Y, Cheng C, Hu Z, Zhuang L, Guo Z, Xie H, Zhang J (2023) Constructed wetlands for pollution control. Nat Rev Earth & Environ 4(4):218–234. https://doi.org/10.1038/s43017-023-00395-z
54. Xu J, Xu Z (2022) China sewage treatment engineering issues assessment. J Clean Prod 377:134391. https://doi.org/10.1016/j.jclepro.2022.134391
55. Xu Z, Xu J, Yin H, Jin W, Li H, He Z (2019) Urban river pollution control in developing countries. Nat Sustain 2(3):158–160. https://doi.org/10.1038/s41893-019-0249-7
56. Yau WK, Radhakrishnan M, Liong S-Y, Zevenbergen C, Pathirana A (2017) Effectiveness of ABC waters design features for runoff quantity control in urban Singapore. Water 9(8):577
57. Yin H, Islam MS, Ju M (2021) Urban river pollution in the densely populated city of Dhaka, Bangladesh: big picture and rehabilitation experience from other developing countries. J Clean Prod 321:129040. https://doi.org/10.1016/j.jclepro.2021.129040
58. Yu Q, Olesen SW, Duvallet C, Grad YH (2024) Assessment of sewer connectivity in the United States and its implications for equity in wastewater-based epidemiology. PLOS Glob Public Health 4(4):e0003039. https://doi.org/10.1371/journal.pgph.0003039
59. Yuan X, Zhang X, Yang Y, Li X, Xing X, Zuo J (2024) Emission of greenhouse gases from sewer networks: field assessment and isotopic characterization. Front Environ Sci Eng 18(10):119. https://doi.org/10.1007/s11783-024-1879-1
60. Zhang D, Gersberg RM, Ng WJ, Tan SK (2017) Conventional and decentralized urban stormwater management: a comparison through case studies of Singapore and Berlin, Germany. Urban Water J 14(2):113–124. https://doi.org/10.1080/1573062X.2015.1076488
61. Zheng M, Zuo Z, Zhang Y, Cui Y, Dong Q, Liu Y, Huang X, Yuan Z (2017) Nitrite production from urine for sulfide control in sewers. Water Res 122:447–454. https://doi.org/10.1016/j.watres.2017.05.048
62. Zheng X, Zou Y, Lounsbury AW, Wang C, Wang R (2021) Green roofs for stormwater runoff retention: a global quantitative synthesis of the performance. Resour, Conserv Recycl 170:105577. https://doi.org/10.1016/j.resconrec.2021.105577

Sustainable Water Management in the Built Environment: Ground Cover Plants as a Solution

Salman Shooshtarian, Motahareh Soheilipoor, Ali Tehranifar, Hassan Salehi, and Peter S. P. Wong

1 Introduction

Access to natural and green environments is a fundamental psychological need for individuals living in urban settings, providing essential opportunities for relaxation, recreation, and mental restoration during their leisure time [49]. Urban green space refers to sections of a city's open areas where land—either naturally occurring or deliberately developed—is covered with vegetation such as trees, shrubs, lawns, and groundcover plants. When properly managed and maintained, these spaces play a critical role in supporting urban ecology, enhancing biodiversity, and promoting the health and well-being of city residents [30].

Globally, the availability of green space per capita varies significantly, ranging from as little as 5 m^2 to as much as 50 m^2, depending on geographic, economic, and urban planning contexts [24]. Water scarcity and declining water quality have become pressing global concerns. The rapid growth of urban populations, coupled with diminishing water resources, is creating a widening gap between current trends and future needs in many regions [14, 16]. As urban populations grow—and as people continue to migrate from rural areas and smaller towns to major metropolitan centres—water consumption rises sharply, placing additional pressure on already

S. Shooshtarian (✉) · P. S. P. Wong
School of Property, Construction and Project Management, RMIT University, Melbourne, Australia
e-mail: salman.shooshtarian@rmit.edu.au

M. Soheilipoor
Faculty of Architecture and Urban Planning, University of Art, Tehran, Iran

A. Tehranifar
Department of Horticultural Science and Landscape Architecture, Ferdowsi University of Mashhad, Mashhad, Iran

H. Salehi
Department of Horticultural Science, School of Agriculture, Shiraz University, Shiraz, Iran

A. Cheshmehzangi et al. (eds.), *Nature-based and Climate Responsive Design for Urban Systems*, Urban Sustainability, https://doi.org/10.1007/978-981-92-2070-0_5

strained urban water systems [37]. For instance, in the semi-arid regions of the southwestern United States, urban water use rises by 40 to 60 percent in the summertime, largely due to irrigating green spaces [21].

Over the past two decades, water-efficient urban greening—commonly referred to as xeriscaping—has been strategically implemented to enhance resource efficiency within the landscape industry. The term 'xeriscaping,' also known as 'dry landscaping,' was originally coined by urban planners as a response to growing water scarcity in metropolitan areas [25]. Xeriscaping aims to minimise the need for intensive agricultural inputs in the creation and maintenance of urban green spaces. This approach offers a sustainable alternative for designing outdoor environments by reducing dependence on municipal water supplies and avoiding the use of non-native or unsuitable plant species [3]. Despite challenging environmental conditions—such as poor soil quality and scarce water resources—xeriscaping has demonstrated considerable success, particularly in arid and semi-arid regions like Australia and Iran [2, 6, 48]. Xeriscaping is considered a complementary approach to the circular economy within the landscape industry, as both emphasise resource efficiency. The seven principles of xeriscaping closely align with the core objectives of circular economy practices as outlined in Table 1.

Urban turfgrasses are among the most extensively utilised plant types in establishing urban green spaces [26]. Most turfgrass species require relatively fertile, well-aerated soils with good drainage and a slightly acidic to neutral pH level to

Table 1 Comparison analysis of principles of circular economy and xeriscaping

Xeriscaping	Circular economy	Similarity
Planning and design	Rethink	Both emphasise re-evaluating traditional approaches to resource use and land development
Soil improvement	Refurbish/Repair	Both focus on improving existing conditions (soil or products) to extend functionality
Efficient irrigation	Reduce	Aims to use fewer resources (water/materials) by optimising systems and technologies
Plant selection	Refuse/Rethink	Encourages the use of native or climate-suitable species, rethinking choices that demand high input
Mulching	Recover/Reuse	Uses organic materials to conserve moisture and return nutrients, mirroring reuse principles
Turf alternatives	Refuse/Reduce	Avoids resource-intensive lawn choices in favour of low-input alternatives
Maintenance	Repair/Repurpose/Refurbish	Encourages ongoing care to preserve functionality, akin to extending a product's lifespan

Source Authors

thrive [27]. However, securing land that meets all these conditions is not always feasible, particularly in urban or degraded environments [27, 28]. In the landscape industry, turfgrasses are often regarded as high-maintenance vegetation due to their substantial water requirements. As such, in arid or water-scarce regions, it is advisable to minimise turf coverage to promote more sustainable and resource-efficient landscape practices [50, 51].

Ground cover plants (GCPs) offer a practical alternative to partially replace turfgrass, as they typically require less water and are easier to maintain. These plants offer a range of benefits that contribute to the development of sustainable urban green spaces. These plant species are particularly valued in landscape design for their ability to introduce a variety of colours and textures [42]. GCPs are typically fast-growing species capable of achieving a lateral spread of up to one meter, making them effective for quickly covering bare soil and enhancing visual appeal in landscape settings [45]. These plants can thrive in challenging conditions where turfgrass typically struggles, such as on slopes, in fully shaded areas, in locations with intense competition from tree roots for water and nutrients, and in soils that are either consistently wet or excessively dry [5] (Table 2).

Another key advantage of GCPs is their natural ability to suppress weeds. They achieve this by producing allelopathic chemicals that inhibit weed growth, limit sunlight penetration to the soil surface, and prevent weed seed germination [41].

Table 2 A summary of the advantages of GPC compared to conventional turfgrasses

Benefit	Description	References
Ability to suppress weeds	GCPs achieve this by producing allelopathic chemicals that inhibit weed growth, limit sunlight penetration to the soil surface, and prevent weed seed germination	[23, 40]
Ability to thrive in challenging, growing conditions	These plants thrive in conditions where turfgrass struggles, including slopes, shaded areas, root-competitive zones, and soils that are overly wet or dry	[5]
Fewer water maintenance requirements	GCPs typically require less water and are easier to maintain	[41]
Quick coverage of bare soil	GCPs are fast-spreading species, often covering up to one meter laterally, making them ideal for quickly covering bare soil and improving visual appeal in landscapes	[45]
Resistance to extreme temperature values	Certain GCPs can adapt to a wide range of climates—from sub-zero winter climates to temperatures above 40 °C	[43]
Resistance to saline soil and irrigation water	GCPs generally show better adaptability to saline soil and irrigation water	[10]
A variety of colours and textures	GCPs are valued in landscape design for their ability to introduce a variety of colours and textures	[31, 42]

Source Authors

These inhibitory substances are released into the environment through various pathways, including volatilisation from leaf tissues, leaching of non-volatile compounds during rainfall, root exudation, and the decomposition of plant residues facilitated by soil microorganisms [23, 40]. In a study examining the weed-suppressing effects of ground cover plants, [11] identified 22 bioactive compounds extracted from lilyturf—an ornamental ground cover commonly used in green spaces—that effectively inhibited the growth of various weed species.

Salt stress is a major environmental constraint that can emerge from various factors, including drought, and often results in stunted or halted plant growth [29]. However, plant species exhibit varying degrees of tolerance to salinity, with some demonstrating greater resilience during the establishment of green spaces (Morales, 2001). Compared to turfgrasses, GCPs generally show better adaptability to saline soil and irrigation water. Their use in landscape design can therefore help mitigate the adverse effects of salinity stress [10].

In recent years, several initiatives have emerged to separate irrigation water used in green spaces from the main drinking water supply—an important step toward achieving more sustainable and efficient water management in urban areas [20]. To date, numerous studies have investigated the use of treated wastewater for irrigating various plant species [33, 36, 42]. Research conducted in this space has highlighted that plant responses to this water source vary significantly [42]. Depending on the species and prevailing environmental conditions, treated wastewater can result in either positive or negative growth outcomes [33, 36]. However, certain GPCs have shown a higher tolerance to treated wastewater compared to other commonly used green space species. Lastly, resistance to extreme temperature values is another notable trait of GCPs. This trait enables them to adapt to a wide range of climates—from sub-zero winter climates to scorching temperatures above 40 °C [43].

2 Literature Review

In Iran, the defined standard green space per person is 12 m^2 [1]. However, none of the country's major cities are currently able to expand their green spaces to meet ideal global benchmarks, largely due to severe water shortages, which remain a major constraint on green space development [7, 35]. Mashhad is Iran's second-largest city and the world's most prominent religious metropolis. It ranks as the second most populous city in the country, with a population of approximately 3 million people and a total area of around 470 km^2. Home to the Holy Shrine of Imam Reza, Mashhad attracts over 20 million tourists and pilgrims each year, making it a major hub for religious and cultural tourism [34].

The unregulated influx of people from surrounding towns, villages, and other provinces has long presented significant challenges for Mashhad [44]. In addition to the growing demand for new green spaces to serve both residents and the millions of annual pilgrims, this ongoing population pressure highlights the urgent need to secure reliable water resources to ensure public well-being. The resulting increase in drinking water consumption, combined with a rise in per capita water use for maintaining green spaces, has created serious urban management issues, ultimately pushing water usage beyond the city's established capacity and planning standards [32, 33].

Mashhad, encompassing its surrounding green belts, private gardens, and the Astan Quds gardens, covers a total of 6,168 hectares of green space, equating to 12.15 m^2 of green space per resident (cited in Shabahang et al. [39]). Meeting the water demand for this extensive green space—and managing the annual rise in consumption—has become a major challenge for urban planners and municipal authorities. Notably, a report by Mashhad's Urban Water and Wastewater Organisation indicates that 180 hectares of the city's green space are still being irrigated using municipal water and sewage network connections [32]. The annual cost of supplying water for Mashhad's green spaces amounts to 10.5 billion rials, placing a substantial financial burden on both city management and residents, a challenge that is further intensified by the growing scarcity of water resources [45]. Hence, Mashhad's Urban Parks and Green Spaces Organisation (MUPG) has strategically shifted towards implementing xeriscaping principles, including the replacement of conventional urban turfgrasses with GCPs [19].

3 Methodology

3.1 Study Area

Mashhad has a cold semi-arid climate, classified as *BSk* under the Köppen climate classification [18]. The city is characterised by a broad temperature range, with recorded extremes reaching 41.6 °C in summer and dropping to −28 °C in winter, and an annual average temperature of 13.28 °C [12]. Mashhad experiences approximately 89 frost days each year and receives an average annual precipitation of 241 mm [8].

GCPs were introduced to Mashhad by private companies in 1998, initially for limited use and propagation. With the occurrence of a seven-year drought, water shortages, and rationing in Mashhad [38], along with increasing attention to urban green space expansion, the use of these plants began to rise. Today, more than 50 species of GCPs are produced in Mashhad, whereas in the past, only a few were used for planting in green spaces. In recent years, there has been a growing need to make green space designs more diverse. At the same time, the move away from traditional grass toward using other types of plants has made GCPs more important in urban greening.

In 2006, a botanical garden was established within the Department of Horticultural Science at Ferdowsi University of Mashhad's College of Agriculture. This garden showcased 22 GCP species selected as practical models for application in urban green spaces. By 2008, the integration of these species into Mashhad's public parks

Fig. 1 The small-scale botanical garden curated by the Department of Horticultural Science. *Source* Google Maps [15]

marked a significant milestone in the city's urban landscape development, reflecting a shift toward more sustainable and adaptive planting practices (Fig. 1).

3.2 Data Collection and Analysis

In this study, 17 GCPs planted in Ferdowsi University's small-scale botanical garden were evaluated using a visual assessment method (VAM). VAM is a semi-quantitative data collection method used to assess and record observable characteristics of physical subjects or environments. It involves systematically observing and judging specific features based on predefined criteria, often using checklists, rating scales, or photographic documentation [9]. This method is useful in fields such as environmental studies, landscape architecture, horticulture, and urban planning. Researchers may visually assess plant health, design aesthetics, cleanliness, spatial arrangements, or maintenance levels [17, 46]. Duncan's Multiple Range Test (DMRT) was used as a post hoc analysis following ANOVA to identify significant differences among treatment means. This test enabled pairwise comparisons of plant performance variables and grouped them according to statistically significant differences at the selected significance level (i.e., $p < 0.05$).

The GCPs evaluated included *Carpobrotus acinaciformis*, *Frankenia thymifolia*, *Lysimachia nummularia*, *Rosmarinus officinalis*, *Hedera helix*, *Potentilla verna*, *Vinca minor*, *Festuca ovina*, *Ajuga reptans*, *Lavandula officinalis*, *Sedum spectabile*, *Lobularia maritima*, *Dichondra repens*, *Sasa pygmaea*, *Sagina subulate*, *Senecio cineraria*, and *Cerastium tomentosum*. During the period of data collection, 20 green space experts were recruited to evaluate a series of ecological metrics by assigning scores on a Likert scale from 1 to 3. These metrics include coverage rate, resistance level to challenging growth conditions, usage rate, foot traffic tolerance, trailing/hanging ability, and the average performance of each species.

A critical factor in assessing the suitability of GCPs for urban green spaces is the total cost associated with their establishment, ongoing maintenance, and potential replanting. To address this, a cost analysis was conducted to compare the economic performance of a selected GCP with that of conventional urban turfgrass in Mashhad's green spaces.

4 Results and Discussion

4.1 Cost Analysis of GCPs Application in Urban Green Spaces

As shown in Table 3, the analysis presents a comparative assessment of the annual planting and maintenance costs per square metre for *Frankenia* sp. and a typical 'sport turfgrass' mix commonly used in Mashhad, Iran. The cost data were obtained from various sources, including [45] and verbal market quotations from major plant nurseries and landscaping firms.

As illustrated in Fig. 2, although the initial cost of purchasing *Frankenia* seedlings is approximately 71% higher than that of turfgrass, their overall annual maintenance

Table 3 Annual costs for the establishment and maintenance of sport turfgrass vs. Frankenia per m^2

Plant	Sport turfgrass (1 m^2)		Frankenia (1 m^2)	
Operation	Quantity per year	Cost (Rials)	Quantity per year	Cost (Rials)
Establishment	35–50 g	2100	6 shrubs	3600
Fertilisation	3 times (1 at planting + 2 top-dressing fertilisation)	152	Once at planting	91
Mowing and edging	20	750	–	–
Weeding	4	729	–	546
Aeration	1	61	–	–
Cleaning and purging	365	1898	365	292
Pest and disease control	2	61	–	–
Irrigation	180 cycles Water outlet valve	2610	40 cycles Water outlet valve	560
Total annual cost		8,361		4,529

Source *Authors*

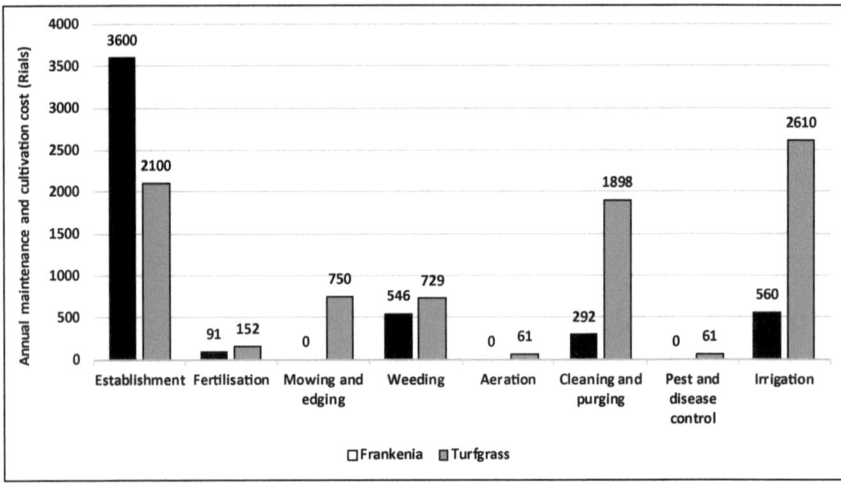

Fig. 2 Comparison of annual cultivation and maintenance costs of turfgrass and Frankenia. *Source* Authors

cost is significantly lower. This difference is primarily due to the intensive horticultural care and higher irrigation requirements needed to maintain turfgrass at an acceptable quality. In contrast, GCPs such as *Frankenia* require minimal maintenance [22]. A key distinction between the two plant types lies in their water consumption: *Frankenia* uses only around 21% of the water required for turfgrass irrigation. Furthermore, turfgrass has greater nutritional needs, typically requiring fertilisation three to five times annually [4], whereas a single application of manure at the time of planting is generally sufficient for *Frankenia.* It has also been reported that the increased use and frequency of fertiliser applications in turfgrass areas contribute to nitrogen leaching into the surrounding environment [4, 47].

As per Table 3, the annual maintenance cost for 100 m^2 of turfgrass is 836,000 rials, whereas replacing it with GCPs reduces the cost to 452,000 rials. This indicates that maintaining turfgrass is approximately 84% more expensive than using GCP alternatives. At present, Mashhad's green spaces require 22.5 million m^3 of water annually for irrigation, yet the city faces a water deficit of 5 million m^3 each year [32]. A substantial portion of the available water is currently allocated to turfgrass irrigation, highlighting the need for more water-efficient urban greening solutions.

4.2 Ecological Performance of Various GCPs

The data were collected with four replications and organised in a randomised complete block design (RCBD). Mean comparisons were performed at a 5% error level using Duncan's multiple range test in the MSTAT-C software [13]. A summary of data analyses is provided in Table 4.

Table 4 Variance analysis (mean squares) of VAM's criteria for the studied GCP species

Sources of variation	DoF	Coverage rate	Resistance level	Usage rate	Foot traffic tolerance	Trailing/ hanging ability	DoF	Mean performance
Replication	2	0/176*	0/176^{ns}	0/ 255**	0/961^{ns}	0/843*	2	5/202**
Species factor	16	1/811**	0/978^{ns}	1/ 706**	2/586**	4/123**	16	1/050^{ns}
Error	32	0/322	0/614	0/338	0/336	0/218	64	0/682

Note *Significant at the 0.05 probability level, **Significant at the 0.01 probability level, and ns: Non-significant. DoF: Degree of Freedom
Source Authors

As shown in Table 5. Frankenia, Sweet Alyssum, and Periwinkle—each with average performance scores of 2.79, 2.39, and 2.53, respectively—demonstrate strong potential as suitable alternatives to turfgrass in urban green spaces. These species showed no significant difference compared to other plants in this regard. Among the evaluated ground cover plants, Pearlwort, Kidney Weed, and Frankenia, with average scores of 3.00, 2.66, and 2.33, respectively, exhibited higher foot traffic tolerance than the other species. However, this difference was statistically significant only in comparison to Fescue.

Additionally, Periwinkle, English Ivy, Giant Pigface, and Sweet Alyssum demonstrated the highest trailing capability among the GPCs, each receiving an average score of 3.00. These species showed no statistically significant difference in trailing performance when compared to Frankenia. Therefore, they are recommended for use on sloped surfaces and in rock gardens, where effective ground coverage and soil stabilisation are essential. In terms of coverage rate, six species demonstrated superior performance, each achieving an average score of 3.00. These plants include Frankenia, Dwarf Bamboo, Spring Cinquefoil, Kidney Weed, Snow-in-Summer, and Giant Pigface. Their coverage performance was significantly higher only when compared to Fescue and Showy Stonecrop.

Concerning resistance, the analysis revealed no statistically significant variation among the evaluated ground cover species. However, Frankenia, Rosemary, and Dusty Miller received the highest resistance scores in this category. Despite their advantages, the GCPs listed in Table 5 generally exhibit lower foot traffic tolerance compared to turfgrass, which is considered a limitation when assessing their suitability as grass alternatives.

Among the evaluated ground cover plants, nine species are flowering plants. These include Snow-in-Summer, Dusty Miller, Lavender, Sweet Alyssum, Showy Stonecrop, Carpet Bugle, Periwinkle, Rosemary, and Giant Pigface. These plants play a valuable role in adding diverse colours to landscapes, particularly in areas where traditional grass only offers a single green hue. Currently, *Dusty Miller*, *English Ivy*, *Rosemary*, and *Lavender* are widely planted in green spaces worldwide. Notably, Frankenia displays seasonal foliage colour changing, appearing green in spring and

Table 5 A summary of the key growth indices of the studied GCPs

Name	Scientific name	Coverage rate	Level of resistance to harsh conditions	Usage rate	Foot traffic tolerance	Hanging ability	Mean
Giant Pigface	*Carpobrotus acinaciformis*	$^{\dagger}3.00^{a}$	2.66^{a}	3.00^{a}	0.00^{g}	3.00^{a}	2.33^{a}
Frankenia	*Frankenia thymifolia*	3.00^{a}	3.00^{a}	3.00^{a}	2.33^{a-c}	$2.66^{a,b}$	2.79^{a}
Moneywort	*Lysimachia nummularia*	2.66^{a}	1.00^{a}	2.00^{a-d}	0.66^{e-g}	$2.00^{b,c}$	1.64^{a}
Rosemary	*Rosmarinus officinalis*	$2.33^{a,b}$	3.00^{a}	3.00^{a}	0.00^{g}	0.00^{g}	1.66^{a}
English Ivy	*Hedera helix*	$2.33^{a,b}$	2.66^{a}	2.33^{a-c}	0.66^{e-g}	3.00^{a}	2.19^{a}
Spring Cinquefoil	*Potentilla verna*	3.00^{a}	2.33^{a}	$2.66^{a,b}$	0.33^{fg}	0.66^{e-g}	1.73^{a}
Periwinkle	*Vinca minor*	2.66^{a}	2.66^{a}	3.00^{a}	1.33^{c-f}	3.00^{a}	2.53^{a}
Fescue	*Festuca ovina*	0.66^{c}	2.33^{a}	$1.33^{c,d}$	2.00^{a-d}	0.00^{g}	1.26^{a}
Carpet Bugle	*Ajuga reptans*	$2.33^{a,b}$	2.00^{a}	1.66^{b-d}	1.33^{c-f}	1.33^{c-e}	1.73^{a}
Lavender	*Lavandula officinalis*	$2.33^{a,b}$	2.66^{a}	3.00^{a}	0.66^{e-g}	0.66^{fg}	1.79^{a}
Showy Stonecrop	*Sedum spectabile*	$1.33^{b,c}$	2.33^{a}	1.33^{a}	0.00^{g}	0.66^{e-g}	1.13^{a}
Sweet Alyssum	*Lobularia maritima*	2.66^{a}	1.66^{a}	3.00^{a}	1.66^{b-e}	3.00^{a}	2.39^{a}
Kidney Weed	*Dichondra repens*	3.00^{a}	2.66^{a}	2.33^{a-c}	$2.66^{a,b}$	$0.33^{f,g}$	2.19^{a}
Dwarf Bamboo	*Sasa pygmaea*	3.00^{a}	2.33^{a}	1.66^{b-d}	1.66^{b-e}	1.00^{d-f}	1.93^{a}
Pearlwort	*Sagina subulata*	0.66^{a}	2.33^{a}	1.00^{a}	3.00^{a}	0.00^{a}	1.39^{a}
Dusty Miller	*Senecio cineraria*	$2.00^{a,b}$	3.00^{a}	2.33^{a-c}	1.00^{d-g}	$2.00^{b,c}$	2.06^{a}
Snow-in-Summer	*Cerastium tomentosum*	3.00^{a}	1.33^{a}	1.00^{d}	1.00^{d-g}	$1.66^{c,d}$	1.59^{a}

†Values in each column sharing the same letter (a–g) are not significantly different as defined in Duncan's test at the 5% statistical level

Source Authors

In Duncan's Multiple Range Test (DMRT), the letters (a, b, c, d, etc.) represent statistical groupings of treatment means. Means that share at least one common letter are not significantly different from each other at the chosen significance level (p 0.05), whereas means with different letters are considered significantly different

Different letters (a–g) indicate significant differences among treatment means according to Duncan's Multiple Range Test (DMRT) at p 0.05. Means sharing the same letter are not significantly different summer, and turning red during autumn and winter. In addition, Rosemary, Lavender, and Sweet Alyssum are characterised by their pleasant fragrance.

It is also worth noting that Rosemary and Lavender were the tallest among the species evaluated. Their height makes them well-suited for use as medium-height hedges to restrict foot traffic, define borders, or act as windbreaks for seasonal flowers and low-growing plants.

5 Conclusions

Resource efficiency has emerged as a critical concern across various industries and sectors, driven by the need to reduce environmental impacts, minimise waste, and optimise the use of materials, water, and energy. In response, numerous strategies and initiatives have been introduced to support industry efforts toward achieving greater efficiency. One of the most widely recognised frameworks in recent years is the circular economy, which promotes systemic change by closing resource loops and designing out waste. This approach offers a powerful, enabling framework for rethinking traditional practices across multiple domains, including the landscape and urban greening sector.

Interestingly, the core principles of the circular economy closely align with those of Xeriscaping—a landscaping philosophy focused on reducing water consumption and improving environmental sustainability. Among the seven foundational principles of Xeriscaping, the replacement of conventional turfgrass with GCPs stands out as a key strategy to improve water efficiency, lower maintenance costs, and enhance ecological performance. This shift not only supports the conservation of valuable resources but also contributes to the creation of greener, more resilient, and circular urban spaces. This study aimed to explore the economic and ecological performance of GCPs through a case study conducted at Ferdowsi University of Mashhad, Iran.

The findings of this study are valuable for a broad range of stakeholders involved in the planning, design, development, and maintenance of urban green spaces. These include landscape architects, urban planners, municipal authorities, environmental consultants, horticulturists, and policymakers. By offering a comparative analysis of the economic and ecological performance of GCPs versus conventional turfgrasses, the study provides a strong evidence base to support more resource-efficient and sustainable planting strategies in urban environments.

For landscape architects and designers, the results, particularly the performance metrics presented in Table 5, offer practical insights into the ornamental and ecological characteristics of different GCP species. This would allow for informed selection and combination of plants based on complementary traits such as coverage rate, foot traffic tolerance, and aesthetic value. Architects and urban designers can draw on these findings to confidently recommend GCPs as viable alternatives to turfgrass, especially in areas where water conservation, biodiversity enhancement, or reduced maintenance is a priority. Urban planners and municipal green space managers can also benefit from this research, as it offers a framework for evaluating long-term cost savings and environmental benefits associated with the integration of GCPs into public green spaces. Horticulturists and nursery operators may use the data to better anticipate market demand for resilient and low-maintenance plant varieties. Additionally, the findings can inform local and regional policy by supporting the development of guidelines and incentive programs that promote the adoption of sustainable landscaping practices.

Future studies are recommended to be conducted to evaluate the adaptability of new or native, cost-effective ground cover species, to better assess their practical potential for application in urban green spaces.

References

1. Alaei N, Mohammadzadeh F, Mostafazadeh R, Talebi Khiavi H (2024) Assessing spatiotemporal urban green space per capita and its connectivity in a small size city in Northwest Iran. GeoJournal 89(4):148
2. Arid Lands Environment Centre (1992) Gardens in desert, imaginative gardening in Arid Australia. Alice Springs, Australia
3. Asadollahi T, Talebi K (2007) The value and application of plants in landscaping of hot and dry areas. Third national conference on green space and urban landscape, Kish, Iran, pp 322–341
4. Barton L, Colmer TD (2006) Irrigation and fertiliser strategies for minimising nitrogen leaching from turfgrass. Agric Water Manag 80(1):160–175. https://doi.org/10.1016/j.agwat.2005.07.011
5. Bowker M, Edinger P (1994) Lawns and ground covers. Sunset Publishing, Menlo Park, California, pp 109–157
6. Bradley P (1994) Garden designed for low water demand. Landscape Australia 16:127–140
7. Dadashpoor H, Jahanzad N (2015) Simulation of future land use changes based on an ecological optimal pattern in Mashhad metropolitan area. Geographical Urban Planning Research (GUPR) 3(3):343–359
8. Deihimfard R, Rahimi-Moghaddam S, Chenu K (2019) Risk assessment of frost damage to sugar beet simulated under cold and semi-arid environments. Int J Biometeorol 63:511–521
9. Dvorak B, Volder A (2010) Green roof vegetation for North American ecoregions: A literature review. Landsc Urban Plan 96(4):197–213. https://doi.org/10.1016/j.landurbplan.2010.04.009
10. Easton CL, Kleindorfer S (2009) Effects of salinity levels and seed mass on germination in Australian species of Frankenia L. (Frankeniaceae). Environ Exp Bot 65(2–3):345–352
11. Eom SH et al (2005) Evaluation of herbaceous perennials as weed suppressive groundcovers for use along roadsides or in landscapes. J Environ Hortic 23:198–203
12. Fallahian A, Bagheri A (2006) Horticulture and green space calendar. Jahad-e-Daneshgahi Publications, Mashhad, Iran
13. Freed R, Einensmith S, Guets S, Reicosky D, Smail V, Wolberg P (1989) User's guide to MSTAT-C analysis of agronomic research experiments. Michigan State University, USA
14. Genhua N, Denise SR (2006) Relative salt tolerance of selected herbaceous perennials and groundcovers. HortScience 110:352–358
15. Google Maps (2025) Ferdowsi University of Mashhad, Botanical Garden, Google Maps
16. Haaland C, van Den Bosch CK (2015) Challenges and strategies for urban green-space planning in cities undergoing densification: a review. Urban For & Urban Green 14(4):760–771
17. He K, Wei L, Li N, Wang B (2022) Construction of urban tree risk assessment and management system based on knowledge graph analysis. For Chem Rev, 1099–1116
18. James PE (1922) Köppen's classification of climates: a review. Mon Weather Rev 50(2):69–72

19. Kazemi F, Jozay M, Salahshoor F, Farhadi H (2021) Application of soil mulches on establishment and growth of native and commercial tall fescue (Festuca arundinacea Schreb.) in an arid environment. Desert 26(1):55–69
20. Keykhosravi R (2007) Estimating the water requirements of green space plants. In: Proceedings of the Third National Conference on Green Space and Urban Landscape, Kish, Iran, pp 84–103
21. Kjelgren RL et al (2000) Water conservation in urban landscapes. HortScience 35:1037–1040
22. Klett JE, Wilson C, Feucht J (2007) Xeriscaping: ground cover plants. Colorado State University Cooperative Extension
23. Kobayashi K (2004) Factors affecting phytotoxic activity of allelochemicals in soil. Weed Biol Manag 4(1):1–7
24. Kwon O-H, Hong I, Yang J, Wohn DY, Jung W-S, Cha M (2021) Urban green space and happiness in developed countries. EPJ Data Sci 10(1):28
25. McErlane R (1989) Water conservation in landscaping. J Prop Manag 54(3):40–42
26. Monteiro JA (2017) Ecosystem services from turfgrass landscapes. Urban For & Urban-Greening 26:151–157
27. Morris KN (2002) A guide to NTEP turfgrass ratings
28. Mortazaeinejad F, Etemadi N (2006) Investigation of salt tolerance in warm-season and cool-season grasses for use in green spaces. J Agric Sci Nat Resour 13:223–313
29. Munns R (2002) Comparative physiology of salt and water stress. Plant, Cell Environ 25:239–250
30. Nakhaie-Moghaddam MA (2007) Investigation of climatic factors limiting the development of urban green spaces in Sistan. In: Proceedings of the Third National Conference on Green Space and Urban Landscape. Kish, Iran, pp 198–204
31. Nameth S, Chatfield J (2001b) Diseases of ground cover plants. The Ohio State Univ. Ext 3064:96–98
32. Nouri H (2009) The municipality: the highest water consumer in Mashhad
33. Rahimizadeh M (2008) The use of treated urban wastewater: a necessity for the sustainable development of Mashhad City. In: Proceedings of the third national congress on recycling and reuse of renewable organic resources in Agriculture. Isfahan, Iran, pp 112–118
34. Razavizadeh N (2023) Mashhad as a pilgrimage destination: toward reinforcing or declining pilgrimage identity? Strat Cult 16(62):61–92
35. Rohollahi I et al (2007) Investigation of the effects of salinity levels on germination and early growth processes of three grass species: Poa pratensis L. and Cynodon dactylon L. In: Proceedings of the third national conference on green space and urban landscape. Kish, Iran, pp 188–197
36. Saadat YA et al (2007) The use of treated wastewater for irrigating tree species and developing urban green spaces. In: Proceedings of the third national conference on green space and urban landscape. Kish, Iran, pp 33–43
37. Salamati Saadatlou MP (1991) Challenges in the maintenance and development of green spaces in large cities. In: Proceedings of the green space seminar, parks and green space organization of tehran municipality. Tehran, Iran, pp 33–49
38. Seyadati Moghaddam MJ, Aghasi A (2004) Quantitative model of Mashhad aquifer. In: Proceedings of the 23rd Geological Sciences Conference of the Geological Survey of Iran. Tehran, Iran, pp 252–259
39. Shabahang S, Gjerde M, Vale B, Balador Z (2019) The problem of lack of green space and rise in surface temperature in the City of Mashhad. In: Sustainability in Energy and Buildings 2018: Proceedings of the 10th International Conference in Sustainability on Energy and Buildings (SEB'18) 10. Springer, pp 258–267
40. Shiraishi S, Watanabe I, Kuno K, Fujii Y (2002) Allelopathic activity of leaching from dry leaves and exudate from roots of ground cover plants assayed on agar. Weed Biol Manag 2(3):133–142
41. Shooshtarian S, Salehi H, Tehranifar A (2011) Study of growth and development features of ten ground cover plants in Kish Island green space in warm season. J Agroecol 3(4):514–524

42. Shooshtarian S, TehraniFar A, Ghani A, Kiani M (2011) Effects of irrigation frequency regimes on morphological and physiological characteristics of six sedum species. Afr J Agric Res 6(26):5694–5700
43. Shooshtarian S (2009) Investigation of the ecological and physiological compatibility of certain groundcover plants in Kish Island. Faculty of Agriculture, Shiraz University
44. Taleshi M (2009) Informal settlements and sustainable urban development: case study of the metropolis of Mashhad in northeastern Iran. World Appl Sci J 7:1235–1238
45. Tehranifar A (2002) Study and investigation of plant species in urban green spaces. Mashhad Parks and Green Spaces Organisation, Mashhad, Iran
46. Tehranifar A, Ghani A, Boghrati M (2011) Comparative study of ornamental potential of six achillea species from Iran. South West J Hortic, Biol Environ 2.
47. Walker WJ, Branham B (2020) Environmental impacts of turfgrass fertilization. Golf Course Manag & Constr, 105–219
48. Walsh K (1993) Water-saving gardening in Australia. Reed Books, Chatswood, NSW, Australia
49. Ward Thompson C, Aspinall PA (2011) Natural environments and their impact on activity, health, and quality of life. Appl Psychol Health Well Being 3(3):230–260
50. Western Australian Water Source Council (1986) Water conservation through good design, Perth
51. Windust A (1995) Drought garden: Management and design for plant survival and your enjoyment. Allscape, Mandurang, VIC

Urban Water in the Age of Climate Crisis: Potentials of Water Sensitive Urban Design Strategies for Gopalganj City, Bangladesh

Tanha Tabassum Tisha, Sourav Zaman, Sourav Ghosh, Efrit Elahi, and Jannatul Ferdousi

1 Introduction

Urban developments, along with rapid urbanization, have significant impacts on the land and water environments. The land development industry often interprets the triple bottom line philosophy in a narrow sense, prioritizing financial viability and profitability over social and environmental considerations. Consequently, while urbanization has contributed to economic growth, it has also accelerated environmental degradation. The impacts of this process vary across regions; however, in developing countries, the pace of modernization, urban expansion, and industrialization has intensified environmental challenges. According to the Asia–Pacific Forum for Environment and Development, rapid development at social and economic level in Asia has caused lack of infrastructure, lack of housing, traffic congestion, and overall decay in environment, especially in the sustainable development of major cities and towns [14]. Key environmental risks associated with urban growth include inadequate access to clean water, rapid increase in population, and artificial methods of cultivation. Furthermore, water resources are increasingly polluted by sewage, industrial effluent, agricultural runoff, and saline intrusion. In many regions, freshwater availability has already become scarce, and projections suggest that with continued population growth, urbanization, and climate change, this scarcity will intensify in the coming century [16].

T. T. Tisha · S. Zaman (✉) · E. Elahi · J. Ferdousi
Department of Architecture, Gopalganj Science and Technology University, Gopalganj, Bangladesh
e-mail: souravzaman.arch@gstu.edu.bd

S. Ghosh
Department of Architecture , Pabna University of Science and Technology, Pabna, Bangladesh

A. Cheshmehzangi et al. (eds.), *Nature-based and Climate Responsive Design for Urban Systems*, Urban Sustainability, https://doi.org/10.1007/978-981-92-2070-0_6

Globally, water scarcity poses a critical challenge, with an estimated 1.6 billion people experiencing economic water shortages and two-thirds of the world's population facing water scarcity for at least one month each year [8, 21]. It was predicted by the United Nations world water development in report 4 that by 2025 more than 1.8 billion people worldwide may face a total water shortage [36]. In developing nations, water-related illnesses account for around 80% of all illnesses and more than one-third of all fatalities. In developing countries, approximately 80% of illnesses and over one-third of deaths are linked to water-related diseases [20], while globally, nearly seven million people die annually due to unsafe water and waterborne diseases [35]. To address this crisis, the United Nations introduced Sustainable Development Goal (SDG) 6.1, which aims to "widespread and equal availability of safe and affordable drinking water for all by 2030" [31]. In Bangladesh, a riparian developing nation in South Asia, more than 97% of the population has access to improved water sources [29]. However, inadequately managed water services serve less than 40% of the population. Previously, water quality concerns were often overlooked, as fresh water was considered abundant and readily available [15]. Today, with a population exceeding 157 million, ensuring safe drinking water has become increasingly challenging without urgent interventions. Achieving SDG 6 in Bangladesh, therefore, requires special attention to hydro-geologically vulnerable and hard-to-reach water-insecure areas [13].

In the southern and southwestern parts of Bangladesh, Sariatpur-Gopalganj region became the gateway for development activities with the completion of the proposed Padma Bridge on the Ganges River. The main force of urban development is booming in the city center of Gopalganj city, located at the bank of the Kaliganga River. Gopalganj City has a canal from the city's origin, piercing the heart of the city while dividing the city into two parts. The canal is 8.44 km long canal in which 7 km contributes to Gopalganj Urban area. The main city growth is along with the canal as the main road of the city follows the canal. However, in the absence of a comprehensive master plan, coupled with rapid urbanization and limited public awareness, both the canal and the city are facing serious environmental and infrastructural degradation. Unplanned development, canal encroachment, and waste dumping are blocking the flow of the canal and disconnecting it from the Kaliganga River, resulting in a stagnant waterbody in the heart of the city. The ignorance of river–canal systems in urban planning has also physically and functionally divided the western and eastern parts of the city. Consequently, the city center is becoming chaotic and overcrowded with traffic congestion and degraded urban environment. Moreover, the city's water resources are increasingly under threat due to unplanned urban growth, encroachment on riverbanks, sedimentation, and pollution. The natural flow of the river has been significantly disrupted, reducing its capacity to manage stormwater and exacerbating the city's flooding issues. Furthermore, the city's drainage systems are poorly maintained, contributing to regular waterlogging and limiting the availability of clean, potable water for its residents.

Along with it, Gopalganj district holds significant importance in terms of water quality risk in Bangladesh's central south. Gopalganj is not particularly close to any coasts, yet the surface water and groundwater systems have been reported of high salinity [28]. Several important variables including stream flow, poor watershed management, climate variability, and human-induced activities are contributing to the decline in groundwater quality in Gopalganj [26]. Contamination has been increasing in surface water and groundwater [27]. Every day, the amount of quality fresh water decreases notably due to contamination of drinking water. As a result, people use commercially supplied jar water in the southern part of Bangladesh as well as in the Gopalganj municipal area.

2 Background Knowledge

Addressing the environmental imbalances created by conventional urban development requires a fundamental shift in planning and design philosophies. A more holistic and multi-disciplined approach is required to ensure planning and design decisions are made with a complete understanding of the environmental, social, and economic consequences. Several developed countries, including Australia, Singapore, France, the United Kingdom, and the United States, have adopted "Water Sensitive Urban Design (WSUD)" as a comprehensive framework for integrating sustainable water management into urban development. Initially emerging from stormwater management practices, WSUD has since evolved into a broader paradigm that emphasizes the holistic management of the entire urban water cycle while embedding it within urban planning and design processes. The original conceptualization of Water Sensitive Urban Design (WSUD) in Perth, Western Australia, emerged as an alternative planning and design framework aimed at reducing urban reliance on large-scale, centralized water infrastructure. Instead, it emphasized the integrated management of all water streams as valuable resources, the promotion of water reuse and recycling, and the mitigation of stormwater impacts on urban water environments. This was to be achieved primarily through strategies such as at-source detention and retention of stormwater, often facilitated by landscaped features incorporated within the urban fabric. Wong and Ashley [32] state that the term WSUD "comprises two parts—'Water Sensitive' and 'Urban Design'." WSUD brings "sensitivity to water" into urban design. The approach seeks to position water as a key element in shaping urban planning and design decisions.

The inter-government agreement on a National Water Initiative defines WSUD as "the integration of urban planning with the management, protection and conservation of the urban water cycle that ensures urban water management as sensitive to natural hydrological and ecological processes." Previous research studies have explored various aspects of urban water management, including the implementation of WSUD in large cities like Dhaka (25, 37], but there is a significant research gap regarding the application of these strategies in mid-sized cities in Bangladesh. Furthermore, there is a lack of comprehensive studies that investigate the integration of spatial planning, community perspectives, and water management in the local context, making it difficult to develop actionable frameworks for urban water resilience. As previously noted, Gopalganj City is experiencing unplanned urban development, leading to the deterioration of groundwater quality and a decline in overall water resources. In response, the research seeks to assess WSUD's viability as a proper planning approach with a sustainable urban water management strategy for mid-sized cities like Gopalganj. These strategies will address the three key management objectives of mains potable water conservation, minimization of wastewater discharge and water quantity and quality management of Gopalganj city of Bangladesh. WSUD Framework for Gopalganj will not only mitigate its current water-related issues but also adopt a long-term, climate-resilient strategy. This approach would enhance the city's water quality, reduce flood risks, and improve the overall livability of the urban environment, thereby contributing to the broader goal of sustainable urban development in Bangladesh and the South Asian region.

2.1 Principles and Global Perspectives of WSUD

Water sensitive urban design (WSUD) is an interdisciplinary planning approach that integrates the entire urban water cycle into city design. According to [1], WSUD aims to "bring sensitivity to water" by giving water due prominence in urban design, ensuring that stormwater and wastewater are treated as resources rather than nuisances. This is the notion of "living with water" that not only uses but also captures, cleans and reuses water locally, reduces flooding and urban heat, and supports biodiversity [30]. Key WSUD principles include mimicking natural hydrological processes, preserving waterways and groundwater, enhancing water quality, and providing ecosystem services (e.g., green roofs, wetlands, permeable pavements). For example, [33] describe a "water sensitive city" built on three pillars: (i) Cities as Water Supply Catchments: access to a diversity of water sources underpinned by a diversity of centralized and decentralized infrastructure; (ii) Cities Providing Ecosystem Services: provision of ecosystem services for the built and natural environment; and (iii) Cities Comprising Water Sensitive Communities: socio-political capital for sustainability and water sensitive decision making and behaviors as illustrated in Fig. 1.

WSUD is closely related to other sustainable urban water frameworks. It overlaps with sponge city concepts (notably in China), low-impact development (LID), sustainable urban drainage systems (SuDS), and blue–green infrastructure approaches. These all share a philosophy of "activating natural processes" and distributing water management across the landscape. Recent work even proposes extending WSUD to Water Sensitive Regional Design (WSRD), integrating hydrological, ecological, and socio-economic factors at landscape scales. WSUD is often misunderstood as a means to control runoff water or solution to waterlogging. On the contrary, WSUD is more than stormwater control, it is a holistic process linking engineering, ecology, and design to achieve water sensitive, resilient cities. Moreover, strategic urban planning is seen as an essential element as the theory requires a comprehensive, long-term framework that embeds water management in land-use plans and policies. However, for micro-level contribution, decentralized solutions

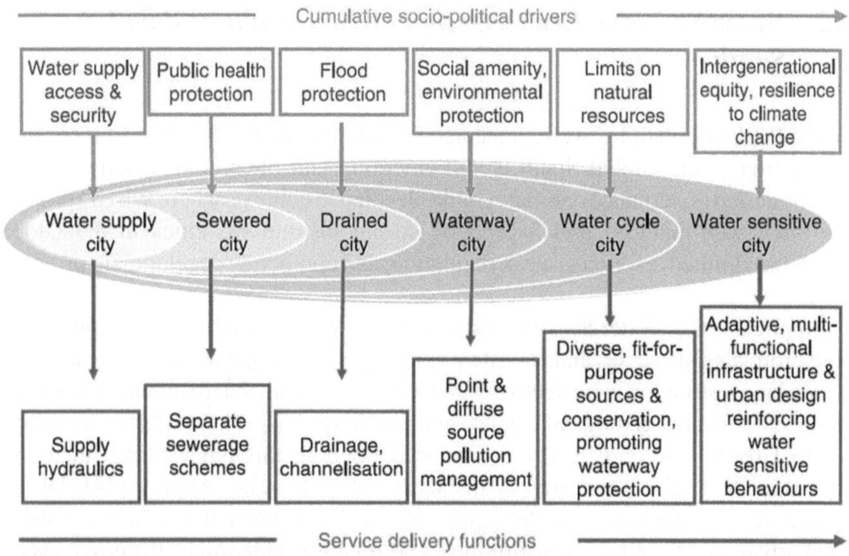

Fig. 1 Positioning water sensitive city. *Source* Adapted from [33]

such as rainwater harvesting, bioretention are emphasized to improve resilience and flexibility.

Originated in Australia through high-value initiatives like "*Cities as Water Supply Catchments*" and the "*Cooperative Research Centre for Water Sensitive Cities,*" WSUD has since influenced cities worldwide. For instance, Singapore has begun implementing WSUD principles and has led regional "*water-sensitive city*" thinking. In Europe, the UK and Netherlands are adopting blue–green infrastructure and SuDS to retrofit drainage and improve water quality. China's "*sponge city*" program applies WSUD-like techniques involving green roofs, rain gardens, and restored water bodies to absorb floods and reduce heat. Even in temperate cities like Philadelphia (USA) or Melbourne (Australia), green infrastructure projects are valued for multibillion-dollar benefits involving flood reduction, urban cooling, etc. Notably, [30] find that in practice, WSUD projects in Australia, Europe and sub-Saharan Africa have yielded lessons on combining water supply, sanitation and drainage management in cities. However, [10] caution that despite strong theoretical support, WSUD often falls short on long-term monitoring, socio-economic outcomes, and policy integration. Nevertheless, researchers call for interdisciplinary collaboration, better community engagement, shared resources and governance reforms to fully functionalize WSUD as a core of climate-resilient urbanism.

2.2 Water Sensitive Urban Design (WSUD) in the Global South

While WSUD principles were pioneered in Australia and adopted in Europe, their application in the Global South is still emerging and often constrained by socio-economic and governance realities. Unlike developed contexts where WSUD projects are often well-financed and backed by strong institutions, cities in the Global South face limited resources, weak regulation, and competing development priorities [4, 10].

Applications of Water Sensitive Urban Design (WSUD) in the Global South reveal both promising successes and persistent challenges. In Bangalore, India, community-driven lake restoration has effectively integrated decentralized treatment systems and wetlands, restoring ecological balance and public spaces [23]. Similarly, Colombo's Metro Urban Development Project reduced flooding through wetland rehabilitation, though it raised concerns of displacement and social inequity [34]. Dhaka, by contrast, demonstrates the difficulties of WSUD implementation where unregulated urban growth and weak institutional coordination undermine efforts to preserve canals and wetlands, resulting in recurring waterlogging. Kathmandu's rainwater harvesting mandates reflect innovative policy adoption, yet enforcement remains inconsistent across socio-economic groups [18]. Beyond South Asia, Medellín's Green Corridors project exemplifies how WSUD-aligned strategies can deliver multiple co-benefits, including urban cooling, flood control, and ecological restoration. Collectively, these cases highlight that while WSUD can enhance resilience, its success depends on governance, community participation, equity considerations, and integration into broader urban strategies.

2.3 Constraints to Implementing WSUD in the Global South

Though WSUD promises resilient urban water management, cities in the Global South face severe constraints in realizing its potential. Key institutional, socio-economic, and infrastructure challenges often limit WSUD outcomes [17, 19]. For example, inadequate governance and fragmented agency roles undermine policy enforcement. However, Socio-economic obstacles are equally daunting. WSUD features like rain gardens or infiltration basin often require funding and public awareness that are scarce in low-income contexts. High upfront costs, limited technical capacity, and low public contribution can stall projects. For instance, Chennai and Bengaluru's utilities mandating household rainwater tanks saw slow adoption because many residents lacked roof space and balked at installation costs. Experts there argue for "community-level systems" instead of owner-built schemes, reflecting the challenge of individual affordability. More broadly, research finds that without strong incentives or subsidies, WSUD is resisted by developers and under-prioritized by cash-strapped cities. Cultural factors also play a role in making

WSUD successful. In Chiang Mai (Thailand), a pilot WSUD design initiative found many residents distrustful of engagement, fearing political misuse of participation or feeling powerless to influence city plans [10]. Such social and political dynamics can leave communities disengaged, undermining maintenance and uptake of green infrastructure.

Infrastructural and land-use constraints further limit WSUD in dense southern cities. High urban densities and impervious surfaces leave little room for green space or infiltration facilities. Studies in Brazil emphasize that "lack of available space for [WSUD] structures" and the absence of integrated planning are critical bottlenecks. In practice, narrow streets and informal housing squeeze out swales and rain gardens. Legacy grey networks also bias engineers toward big drains: as one Lucknow (India) study noted, the city "depends on its conventional and centralized stormwater drains" but "in the absence of a robust stormwater management policy, its negative outcome is observed through a failure of the drainage systems" [2]. These barriers have shaped many project outcomes. Chennai's WSUD mandate, once heralded as visionary, "has faltered" in practice due to lax enforcement and inadequate public education. Bengaluru and Delhi continue to face severe water stress despite policy reforms, because technical interventions were unevenly applied and poorly monitored. In contrast, even successful pilots like China's "sponge cities" rely on strong state investment and planning capacity seldom found in poorer cities. The cumulative effect is that WSUD projects in the Global South often achieve only partial success: they may improve local awareness or provide limited flood relief, but rarely transform city-wide resilience [19]. In summary, institutional fragmentation, socio-economic deficits, and infrastructural rigidity tend to constrain WSUD's promise in the Global South, undercutting the climate-adaptation objectives that planners aim for.

3 Research Design

The present research adopts a qualitative approach, as its methods and analytical strategies are grounded in the collection and interpretation of qualitative data obtained from stakeholders at multiple levels. Qualitative inquiry is particularly suited to this study because it seeks to understand the contextual settings in which participants experience and articulate urban water challenges. Rather than focusing narrowly on the researcher or participants, qualitative research emphasizes the exploration of the phenomenon itself. In this case, presenting the dynamics of water resource decline and its impacts on medium-sized cities such as Gopalganj is our prime focus.

Given the contextual nature of the inquiry, this study employs a case study methodology, which investigates an issue within a bounded system (i.e., a defined setting or community). The first objective is pursued through a detailed survey of Gopalganj Municipality, combining both primary and secondary data. Primary data will be collected using semi-structured interviews; key informant interviews (KIIs), participant observation, and focus group discussions (FGDs), thereby capturing community perspectives on historical and current water management issues. In addition, planning documents and datasets from local authorities, including master plans, spatial maps, and water resource network maps, will be analyzed to identify critical problem zones requiring spatial interventions.

The second objective focuses on morphological analysis of the city to understand its transformation over time and its implications for water management. Building upon this contextual understanding, the study will then engage with secondary literature on Water Sensitive Urban Design (WSUD). Relevant theories, principles, and case applications of WSUD will be critically reviewed to assess their relevance and adaptability in the Global South context. Successful and unsuccessful cases will be examined to extract lessons for implementation.

Finally, the study aims to synthesize these insights into a strategic WSUD framework tailored for Gopalganj. This framework as illustrated in Fig. 2 is expected to serve as a holistic urban design approach that integrates water sensitivity into local planning practices and addresses context-specific challenges in mid-sized Bangladeshi cities.

4 Urban Water Management Challenges in Gopalganj

Gopalganj Municipality (now 30.7 km^2 after 2024 expansion) has only 9,900 public water connections to serve a dense urban population. Local authorities estimate an average demand around 60 Million Liters per Day (MLD), yet only about 13–14 MLD is delivered which barely covers around 25% of need. The main treatment works (two plants built in 2001 and 2020) has lost capacity which was originally 540 m^3/h, it now functions at only around 300 m^3/h. Even running 24/7, this old plant meets just around 25% of customer requirements. As a result, many wards receive supply only intermittently (often 1–2 h per day). This shortfall forces residents to haul water from distant or unsafe sources. A recent report by a leading newspaper states both groundwater and surface sources in Gopalganj are heavily contaminated [6]. On a KII, a government official of Public Health Engineering (DPHE) in Gopalganj district reported high iron, salinity and arsenic in much of the groundwater. They also expressed their deep concern as the water scarcity rises during summer; even deep tube wells cannot be installed due to high iron, salinity, and arsenic level. A 2024 study of Kashiani (Gopalganj) found that 61% of tube-well samples were "poor quality," with iron and arsenic greatly exceeding WHO limits. Virtually all samples posed *elevated chronic health risks* (100% for children) from arsenic. Locals report widespread waterborne illness (diarrhea, skin diseases) from contaminated ponds and well water (Shaibur et al., 2021).

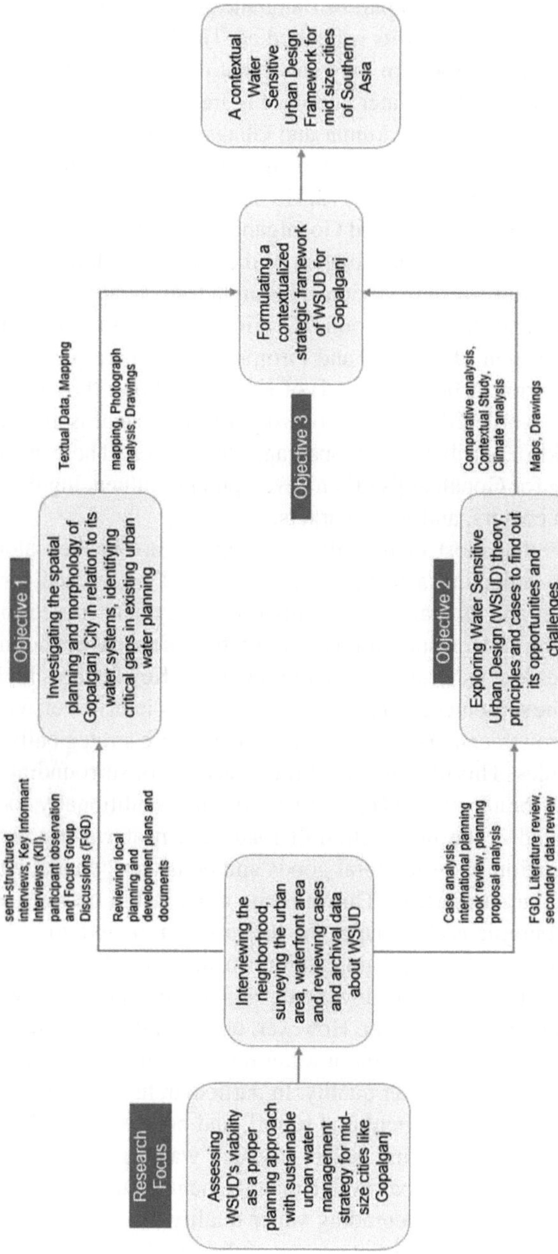

Fig. 2 Analytical framework. *Source* Author

4.1 Geography and Water-Based City Growth

Gopalganj, located in the southern part of Bangladesh, is significantly shaped by its geographical features, particularly its water bodies. The Modhumoti River and Kaliganga River, along with an 8.44 km long canal that divides the city into two parts, play a crucial role in the city's water supply. Figure 3a shows the water network of the Gopalganj district including kumar and Ghagor river, Wabda and Madaripur canal and the canal within Gopalganj city. Gopalganj is a district within Bangladesh's Dhaka Division, with a population of 1,172,415 and covering an area of 1,490 square kilometers. The main town, also named Gopalganj, lies along the banks of the Kaliganga River. The river Kaliganga is a branch of the mighty Brahmaputra. This tidal river, located in the southwestern region of Bangladesh, is approximately 78 km long and 500 m wide and flows year-round. Originating from the Madhumati River, it flows through Gopalganj, Bagerhat, and Pirojpur before merging with the Kacha River in Pirojpur district. A branch (canal) of Kaliganga flows through the heart of the city dividing the whole urban area into two parts. The canal is an 8.44 km long canal, in which 7 km contributes to Gopalganj Urban area. These water systems form the foundation for Gopalganj's urban development, influencing the location of settlements, growth centers, and rural markets.

Figure 3b shows that most of the city's settlements are located along or near these water bodies, reflecting a historical and practical preference for access to water. The main river and its branches not only serve as critical sources of water for domestic, agricultural, and transportation purposes, but also act as economic lifelines, supporting industries like fisheries and facilitating trade. Key features highlighted in the Fig. 3c include the settlement patterns, which are tightly interwoven with the water network, and the growth centers and rural markets that are strategically positioned near these water bodies. This placement optimizes access for surrounding communities and ensures the vibrancy of local economic activities. Additionally, local markets are commonly located at the intersection of roads and near water sources, which aids in the transportation of agricultural goods and enhances connectivity between rural producers and larger markets. This intricate relationship between waterways, settlements, and economic nodes shapes the urban structure and contributes to the sustainable livelihoods of the local population. In the Fig. 3b, it is clear that the branch canal of Kaliganaga River is flowing through the heart of the city connecting 7 local markets and one growth center. However, urbanization, canal encroachment, and pollution have led to deterioration in water quality, with significant impacts on both surface water and groundwater quality. In particular, high salinity and contamination from industrial waste, agricultural runoff, and poor watershed management have exacerbated the issue, making access to clean water a significant challenge. The lack of a comprehensive urban water management system has contributed to stagnation in the canal, further degrading water quality. As from Fig. 3 it is clear that Urbanization overpowered the waterscape of the city. In just 30 years, the water resource has shrunk to its one-fourth.

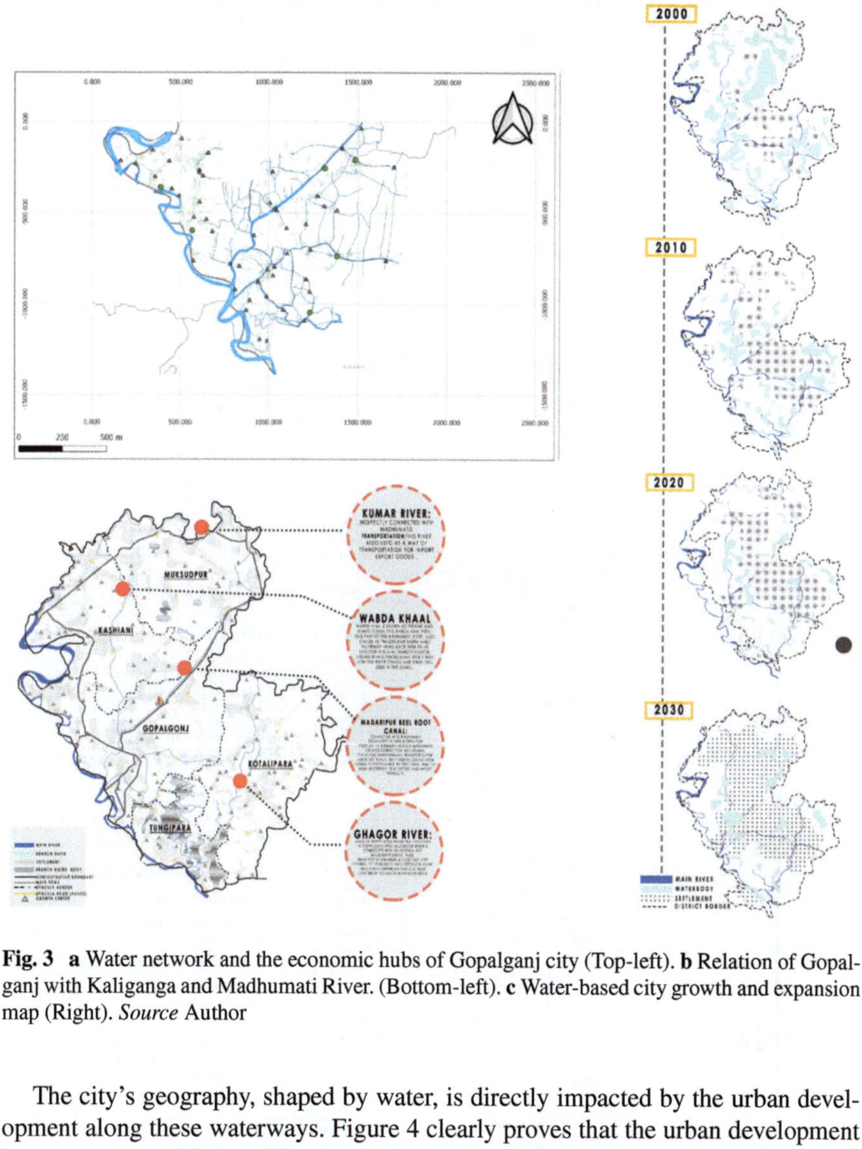

Fig. 3 **a** Water network and the economic hubs of Gopalganj city (Top-left). **b** Relation of Gopalganj with Kaliganga and Madhumati River. (Bottom-left). **c** Water-based city growth and expansion map (Right). *Source* Author

The city's geography, shaped by water, is directly impacted by the urban development along these waterways. Figure 4 clearly proves that the urban development has overgrown with time threatening the water resources and greeneries. In just 40 years the water resource areas are reduced from 70% of total area to 30% of total area. On the contrary, the built area expanded from 10% of total area to 70% of total area. Unplanned urban development and water body negligence caused decline of water resources and the disruption of water network and flow. As water resources become scarcer and the presence of salinity has made people increasingly rely on commercially supplied jar water, highlighting the acute water scarcity in the city.

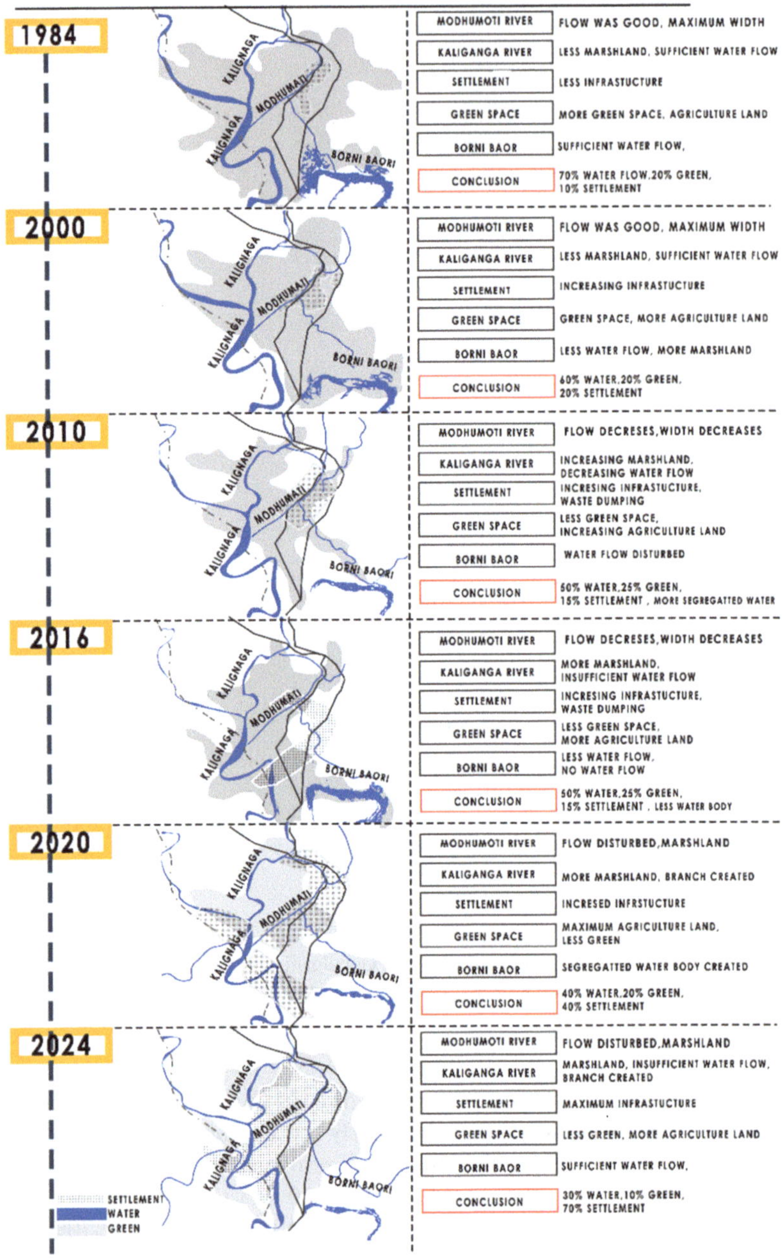

Fig. 4 Morphological transformation of Gopalganj city. *Source* Author

4.2 Drainage and Waste Management Scenario

Gopalganj's stormwater drainage is grossly inadequate. During recent monsoon rains, over 100 families in multiple wards were stranded as streets and even farmlands flooded to waist depth. Locals blame an obsolete drainage network: narrow, unmapped drains blocked by solid waste allow foul, stagnant water to back up onto roads. During interview sessions, the municipality's own administrator notes that unplanned construction in low-lying areas has exacerbated waterlogging, and that without a "modern, well-planned drainage system," even moderate rains cause inundation.

The existing drainage system of Gopalganj is directly connected to the Canal which is from Kaliganga River (See Fig. 5). However, the canal is now almost dead with very poor ecological condition. The entire city's stormwater and wastewater flow into this nearly dead canal, making it even more polluted with dirt and waste. The situation becomes critical when the same river water, after treatment, is used again as a source of drinking water. In this way, the drainage system and the Kaliganga River form a fragile cycle where pollution and water supply are closely linked, creating serious environmental and public health concern. Figure 6b illustrates the current waste management practices and the areas most affected by pollution in both urban and rural zones. The map highlights the red-marked zone as the main city area managed by the municipality, where waste is disposed of in three designated dumping sites. This area is responsible for a significant portion of the city's waste, but it still faces challenges with unregulated dumping practices.

Key findings from the Fig. 6a include red-marked Zone 2, which is heavily polluted due to market waste being directly dumped into the canal, contributing significantly to water pollution. Similarly, Zone 1 also experiences pollution from waste being directly disposed of into the river. Other residential areas use informal methods of waste disposal, with residents discarding waste behind their houses in designated spaces. Additionally, riverside communities are observed to dispose of waste directly into the river, further worsening the environmental degradation. The lack of effective waste management infrastructure in these areas and the absence of municipal coverage in peripheral zones are the main factors leading to water pollution and environmental harm in Gopalganj. The canal front areas, situated at the city's core, are largely accessible due to initiatives undertaken by the development authority, which include the introduction of parks, walkways, and seating zones intended to activate the space. However, these parks provide only limited opportunities for recreation or relaxation. The absence of adequate tree coverage and shaded areas along the walkways and within the parks restricts their usability during daytime hours. Moreover, the lack of child-friendly facilities or play opportunities diminishes their attractiveness to families and children. In the evenings, these spaces are occasionally occupied by groups of young males for social gatherings, though often associated with undesirable activities such as substance use. Apart from this, the walkways which share boundary wall become active zone for criminal activities due to absence of surveillance. Riverside areas are mostly inaccessible for city people. The river can

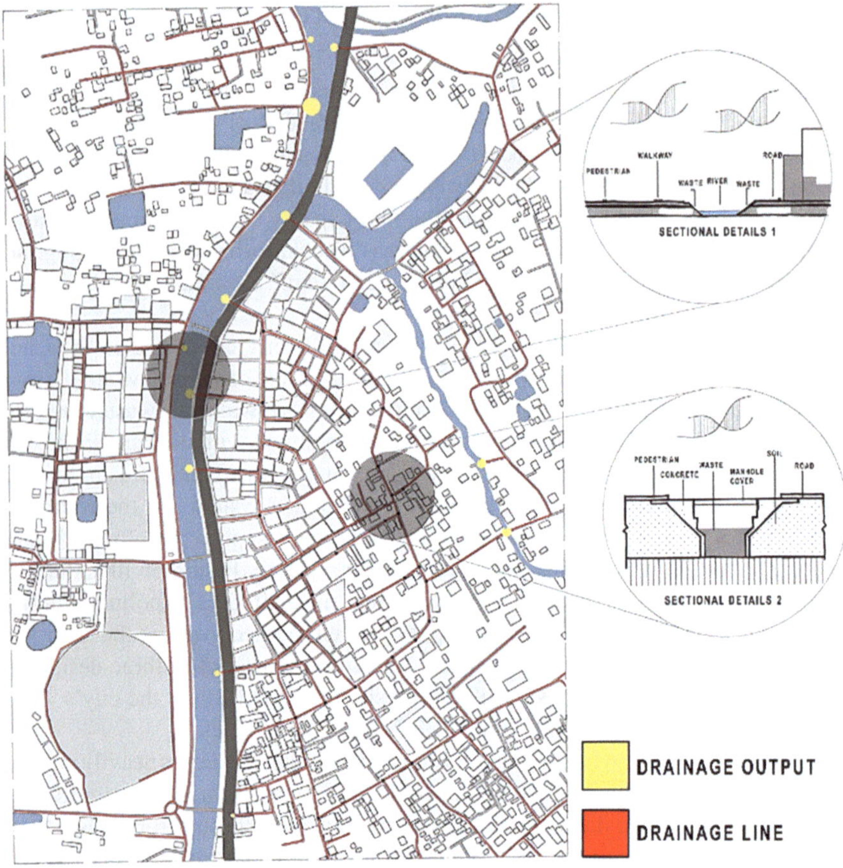

Fig. 5 Drainage network and output. *Source* Author

only be viewed or visited from the Chapail bridge. No measure is taken to make the riverfront accessible.

5 Discussion and Implications

The key opportunity in Gopalganj lies in the integration of green–grey infrastructure as part of a more holistic urban water management strategy. Instead of relying solely on conventional hard-engineered drainage channels, the incorporation of WSUD elements such as bio-retention basins, infiltration swales, and retention ponds can help attenuate peak stormwater flows while simultaneously enhancing groundwater recharge. The results from geographical, morphological, and drainage

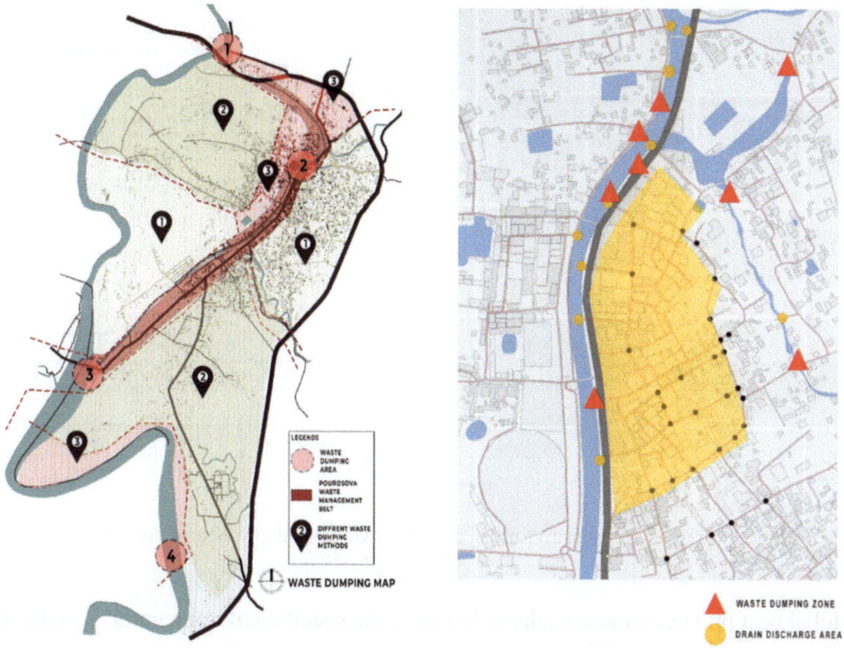

Fig. 6 **a** Waste dumping belt of the city (Left). **b** Micro scale waste dumping zone and drain discharge area at Chowrongi Bazar (Right). *Source* Author

mapping analyses underscore the relevance of the three pillars proposed by [33]—accessibility, provision, and socio-political dynamics—in shaping the feasibility of WSUD interventions. Drawing on insights from case studies, spatial analysis, and a review of relevant literature, complemented by evidence gathered through key informant interviews (KIIs), focus group discussions (FGDs), and stakeholder interviews, this study recommends the formulation of a strategic framework accompanied by context-specific design solutions for Gopalganj.

5.1 Proposed Strategies and Framework for Gopalganj

Here Fig. 7 depicts the Water Sensitive Urban Design (WSUD) framework proposed for Gopalganj aims to address critical water management issues arising from unplanned urban growth, pollution, and water scarcity. Given Gopalganj's unique challenges—such as disconnected and nearly dead water network, encroachment on the Kaliganaga River and canal, inaccessible and inactive waterfront area and deteriorating water quality—this framework integrates sustainable, nature-based solutions designed to manage water resources intact and accessible, enhance water quality, and improve the city's resilience to climate change. The selected techniques draw from

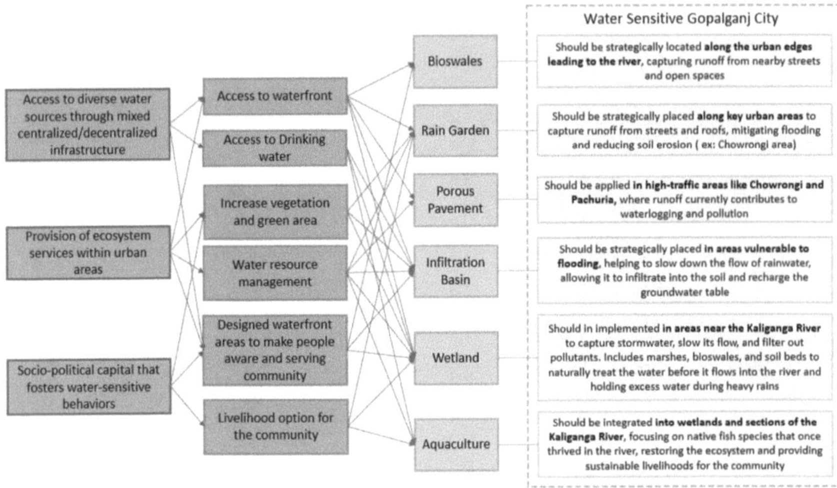

Fig. 7 Framework of water sensitive urban design for Gopalganj city. *Source* Author

global best practices and are adapted to meet the specific environmental, social, and infrastructural needs of Gopalganj.

Based on a comprehensive review of the literature, the proposed WSUD techniques for Gopalganj include bioswales, rain gardens, porous pavements, infiltration basins, wetlands, and aquaculture (See Fig. 8). These strategies have been proven to reduce runoff, promote groundwater recharge, mitigate flooding, and improve the ecological health of urban water systems. The following Table 1 summarizes these techniques, along with the key contributors in the field.

5.2 *Strategic Design Suggestions for Identified Spaces*

Here Fig. 8 outlines the key problem areas and proposes a comprehensive schematic solution designed to optimize overall performance and address underlying issues.

5.2.1 Bioswale

A bioswale is a shallow, vegetated channel designed to capture, slow, and filter stormwater runoff through layers of vegetation, mulch, and engineered soil, thereby improving water quality before it enters larger water bodies [7]. In Gopalganj, bioswales are proposed as a key WSUD intervention to mitigate urban runoff into the Modhumoti River. They are strategically applied in the previously identified problem zones, capturing runoff from nearby streets and open spaces.

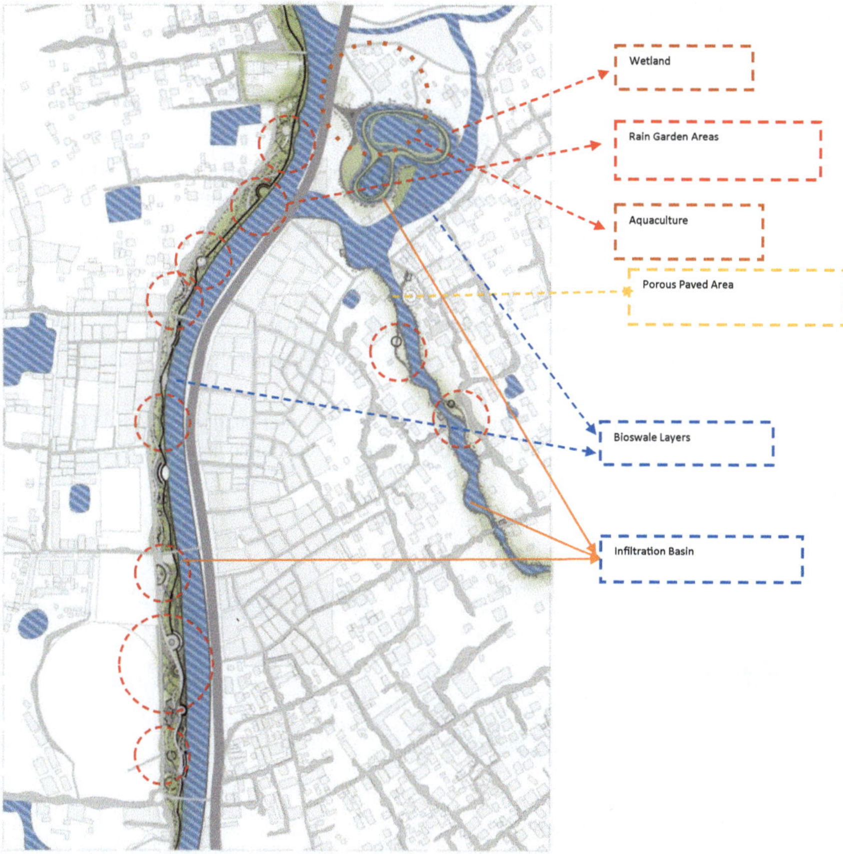

Fig. 8 identified problem zoning and schematic proposed solution. *Source* Author

Figure 9 illustrates the functional layers of the bioswale as applied in these zones, showing how the system operates to manage stormwater effectively. Designed with longitudinal slopes of 1–2% in response to local rainfall patterns, the bioswales are expected to function efficiently during monsoons, reducing polluted inflows while supporting groundwater recharge. In addition, they enhance riverfront restoration by improving ecological health and establishing green corridors that promote community interaction and resilience.

5.2.2 Rain Garden

A rain garden is a shallow, vegetated depression that captures and infiltrates stormwater runoff while filtering pollutants through soil and plants. It helps reduce flooding, soil erosion, and pollution, while also recharging groundwater and

Table 1 Key WSUD techniques

WSUD technique	Description	Key authors
Bioswale	A vegetated channel designed to capture, slow down, and filter stormwater runoff, improving water quality	[7, 32]
Rain Garden	A shallow depression planted with vegetation to capture and infiltrate stormwater, reducing runoff, and filtering pollutants	[7, 12]
Porous Pavement	Pavement that allows water to pass through, reducing runoff and promoting groundwater recharge	[9, 32]
Infiltration Basin	A landscaped area designed to collect and absorb stormwater, reducing flood risks, and recharging groundwater	[3, 5]
Wetland	A natural system that filters water, improves quality, and supports biodiversity	[11, 22]
Aquaculture	Sustainable fish farming that can improve water quality and restore local biodiversity while providing economic benefits	[7, 24]

Source Author

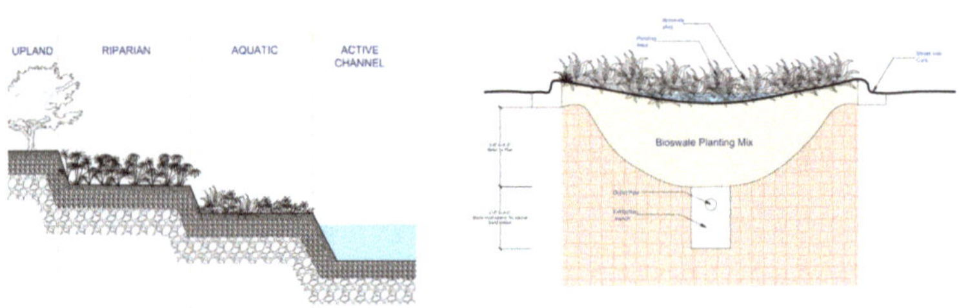

Fig. 9 Bioswale section. *Source* Author

enhancing urban biodiversity [12]. In Gopalganj, particularly in the Chowrongi area, rain gardens are proposed as a WSUD technique to intercept runoff from streets and roofs before it enters the Modhumoti River. Figure 10 illustrates their application, showing strategic placement across key urban zones. Using deep-rooted native vegetation, these gardens are designed to infiltrate water within 24–48 h, preventing stagnation and mosquito breeding. Beyond stormwater management, they will improve water quality, strengthen urban ecosystems, and support the ecological restoration of the Modhumoti River.

5.2.3 Porous Pavement

Porous pavement is a permeable surface that enables stormwater infiltration, thereby reducing runoff, improving groundwater recharge, and filtering pollutants to enhance water quality. It also mitigates flood risks, lowers urban heat, and reduces long-term drainage maintenance needs [9]. In Gopalganj, porous pavements are proposed as a

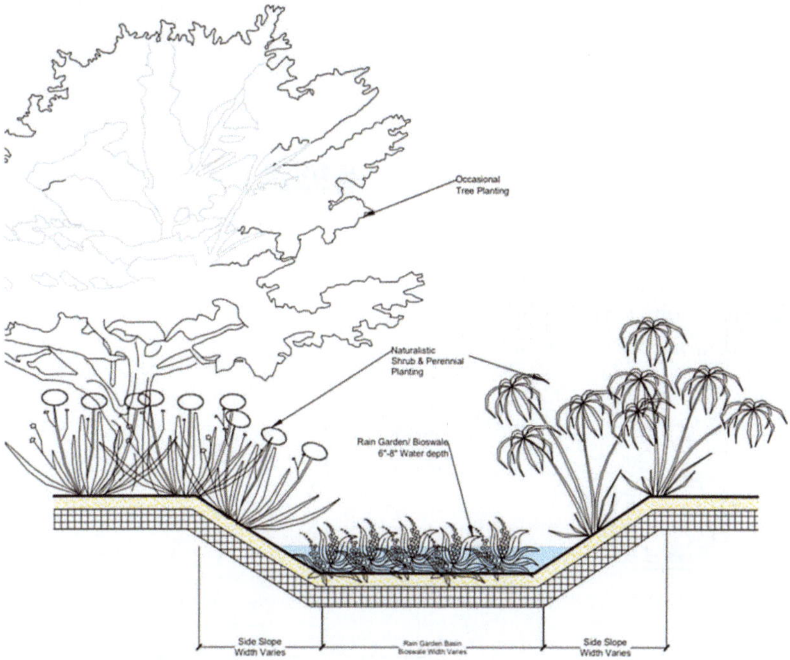

Fig. 10 Rain garden section. *Source* Author

WSUD strategy for high-traffic areas such as Chowrongi and Pachuria, where runoff contributes to waterlogging and pollution. Figure 11 illustrates their application, designed with gentle slopes and shallow depressions to slow water flow and trap sediments before reaching the Modhumoti River. Incorporating native vegetation will further enhance filtration and create green urban spaces. Collectively, this approach eases pressure on the city's drainage network while improving ecological health, flood resilience, and urban livability.

5.2.4 Infiltration Basin

An infiltration basin is a shallow, landscaped feature designed to capture stormwater, slow its flow, and promote natural infiltration into the ground. This process reduces runoff, mitigates flooding, and filters pollutants while recharging groundwater. In Gopalganj, where monsoon rains frequently cause severe waterlogging, infiltration basins are proposed as a key WSUD intervention [3]. Figure 12 illustrates their application in flood-prone urban areas near the Modhumoti River, strategically positioned to intercept runoff before it reaches the river. Designed in line with local rainfall patterns, these basins ensure optimal absorption during the rainy season without overloading the drainage system. Beyond hydrological benefits, infiltration

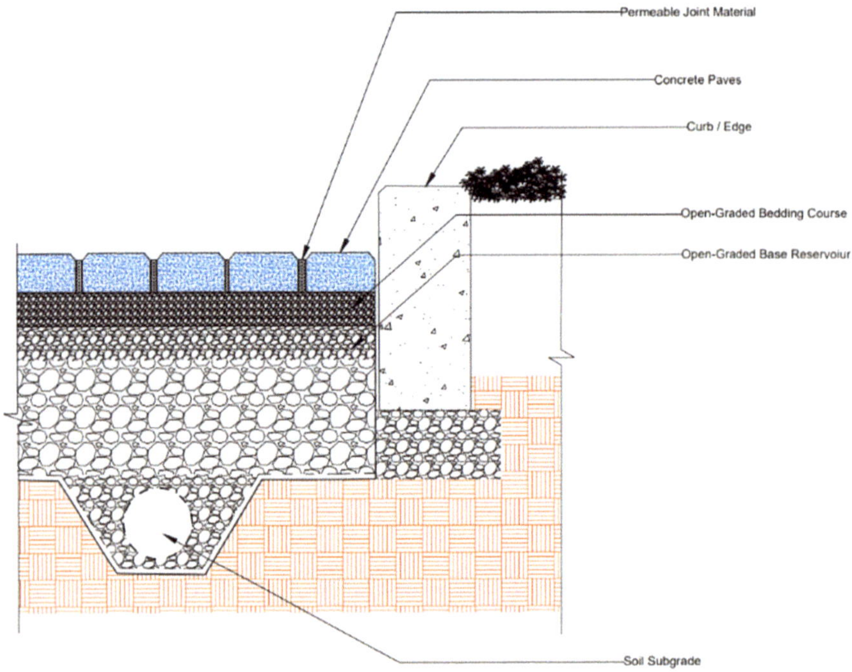

Fig. 11 Porous pavement section. *Source* Author

basins also create multifunctional green spaces, enhancing ecological health, urban aesthetics, and the city's long-term climate resilience.

5.2.5 Wetland and Aquaculture

A wetland is a nature-based solution designed to sustainably manage stormwater and drainage by slowing water flow and naturally filtering pollutants before they reach rivers. In this system, water enters a basin where heavy debris settles and then moves through planted marshes, bioswales, and soil beds that further purify the water [22]. Wetlands also store excess water during heavy rainfall, helping to prevent flooding and recharge groundwater. This design improves water quality, reduces flood risks, and promotes biodiversity. For Gopalganj, a proposed wetland near the Modhumoti River will address drainage and water quality issues by incorporating these features and providing storage for stormwater. The wetland will be populated with native plants to restore biodiversity, attract wildlife, and support a sustainable aquaculture zone, where fish farming will revive local fish populations and the community's cultural and economic ties to the river. Figure 13 illustrates the integration of aquaculture within the wetland, showing how fish farming can support both ecological restoration and local livelihoods. This comprehensive approach,

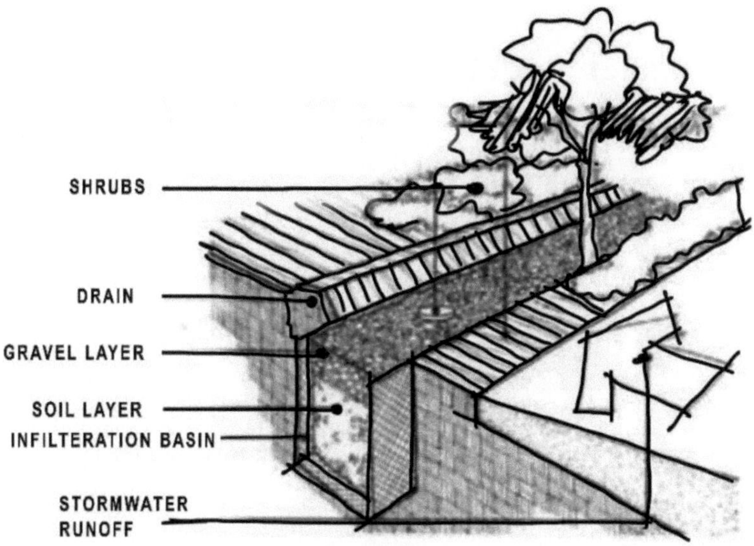

Fig. 12 Infiltration basin section. *Source* Author

combining water management and aquaculture, will contribute to the ecological health of the Modhumoti River while providing social and economic benefits to the Gopalganj community.

6 Concluding Remarks

The study highlights that Gopalganj, like many mid-sized Bangladeshi cities, faces critical challenges of water scarcity, contamination, flooding, and ecological decline linked to rapid urbanization and weak planning. Conventional infrastructure has failed to cope, making Water Sensitive Urban Design (WSUD) a vital pathway for sustainable and resilient urban water management. Adapting global lessons to local realities, the proposed WSUD framework integrates green–grey solutions such as bioswales, rain gardens, porous pavements, infiltration basins, wetlands, and aquaculture. These approaches can revive degraded water systems, manage stormwater, improve water quality, and create multifunctional landscapes that support both resilience and community wellbeing.

However, the transition toward a water sensitive Gopalganj will require more than technical interventions. Effective governance, strong community participation, policy integration, and sustained investment are essential for overcoming socio-economic and institutional barriers. If pursued strategically, WSUD can not only mitigate current water crises but also position Gopalganj as a model for climate-resilient urban development in Bangladesh and the wider Global South.

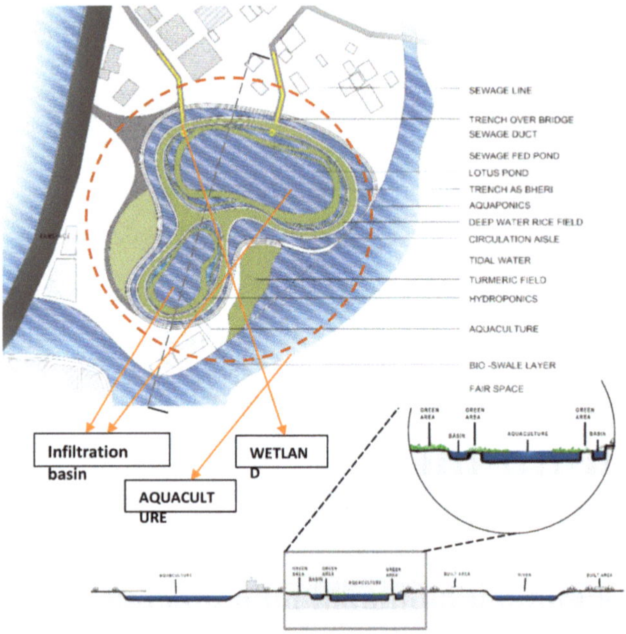

Fig. 13 Wetland and aquaculture zone with section. *Source* Author

References

1. Ashley R, Lundy L, Ward S et al (2013) Water-sensitive urban design: opportunities for the UK. Proceedings of the institution of civil engineers: municipal engineer, 166 (ME2), pp 65–76. ISSN 0965-0903. https://doi.org/10.1680/muen.12.00046
2. Bastia J, Mishra BK, Kumar P (2021) Stormwater infiltration in rapidly urbanizing Lucknow city. Hydrology 8(2):93. https://doi.org/10.3390/hydrology8020093
3. Birch GF, Fazeli MS, Matthai C (2005) Efficiency of an infiltration basin in removing contaminants from urban stormwater. Environ Monit Assess 101(1):23–38
4. Brown RR, Keath N, Wong THF (2009) Urban water management in cities: historical, current and future regimes. Water Sci Technol 59(5):847–855. https://doi.org/10.2166/wst.2009.029
5. Dechesne M, Barraud S, Bardin JP (2004) Spatial distribution of pollution in an urban stormwater infiltration basin. J Contam Hydrol 72(1–4):189–205
6. Dhaka Tribune (2024, May 27) Groundwater contamination triggers pure water crisis in Gopalganj. Dhaka Tribune
7. Dunnett N, Clayden A (2007) Rain gardens. Managing water sustainably in the garden and designed landscape
8. FAO (2007) Coping with water scarcity: challenge of the twenty-first century. Un Water 24:28–29
9. Ferguson B (2005) Porous pavements. CRC Press
10. Ghaderian M, Hakimian P, Shahab S (2025) The trajectory of water sensitive urban design: integrating water management with urban planning and design. Australian Planner 61(1):29–42. https://doi.org/10.1080/07293682.2025.2526201
11. Gibbs JP (2000) Wetland loss and biodiversity conservation. Conserv Biol 14(1):314–317
12. Hlavínek P, Zeleňáková M (2015) Storm water management. Springer, Switzerland. Tillgänglig: https://link.springer.com/content/pdf/bfm%A978-3-319-25835-5%2F1.pdf. 18 Feb 2020

13. Hossain MJ, Chowdhury MA, Jahan S, Zzaman RU, Islam SLU (2021) Drinking water insecurity in southwest coastal Bangladesh: how far to SDG 6.1? Water 13(24):3571
14. Ichimura, M (2003) Urbanization, urban environment and land use: challenges and opportunities. Asia-Pacific Forum for Environment and Development, Expert Meeting
15. Islam M, Shamsad S (2009) Assessment of irrigation water quality of Bogra district in Bangladesh. Bangladesh J Agric Res 34(4):507–608
16. Jackson JB, Kirby MX, Berger WH, Bjorndal KA, Botsford LW, Bourque BJ, Bradbury RH, Cooke R, Erlandson J, Estes JA (2001) Historical overfishing and the recent collapse of coastal ecosystems. Science 293(5530):629–637
17. Khalaji F, Zhang J, Sharma AK (2025) Social and economic impacts of water sensitive urban design: a review. Water 17(1):16. https://doi.org/10.3390/w17010016
18. Khanal G, Maraseni T, Thapa A, Devkota N, Paudel UR, Khanal CK (2023) Managing water scarcity via rainwater harvesting system in Kathmandu Valley, Nepal: People's awareness, implementation challenges and way forward. Environ Dev 46: Article 100850. https://doi.org/10.1016/j.envdev.2023.100850.
19. Kumar A, Button C, Gupta S, Amezaga J (2023) Water sensitive planning for cities in the Global South. Water 15(2):235. https://doi.org/10.3390/w15020235
20. Mazurkiewicz J, Mazur A, Mazur R, Chmielowski K, Czekała W, Janczak D (2020) The process of microbiological remediation of the polluted Słoneczko reservoir in Poland: for reduction of water pollution and nutrients management. Water 12(11):3002
21. Mekonnen MM, Hoekstra AY (2016) Four billion people facing severe water scarcity. Sci Adv 2(2):e1500323
22. Mitsch WJ, Gosselink JG, Zhang L, Anderson CJ (2009) Wetland ecosystems. John Wiley & Sons
23. Nagendra H, Bai X, Brondizio ES, Lwasa S (2019) The urban south and the predicament of global sustainability. Nat Sustain 1:341–349. https://doi.org/10.1038/s41893-018-0101-5
24. Pillay TVR (2008) Aquaculture and the environment. John Wiley & Sons
25. Schulze J, Gehrmann S, Somvanshi A, Rudolph-Cleff A (2024) From district to city scale: The potential of water-sensitive urban design (WSUD). Water 16(4):582
26. Shaibur MR, Howlader M, Ahmmed I, Sarwar S, Hussam A (2024) Water quality index and health risk assessment for heavy metals in groundwater of Kashiani and Kotalipara upazila, Gopalganj, Bangladesh. Appl Water Sci 14(5):106
27. Shaibur MR, Howlader M, Nahar N, Hossain MS, Mamun AM, Shohan MH, Selim A (2024) Application of pollution indices to determine pollution intensities in the groundwater of Gopalganj (south-central part), Bangladesh. Groundw Sustain Dev 26:101206
28. Shammi M, Rahman MM, Islam MA, Bodrud-Doza M, Zahid A, Akter Y, Quaiyum S, Kurasaki M (2017) Spatio-temporal assessment and trend analysis of surface water salinity in the coastal region of Bangladesh. Environ Sci Pollut Res 24(16):14273–14290
29. Supply WUJW, Programme SM (2015). Progress on sanitation and drinking water: 2015 update and MDG assessment. World Health Organization.
30. Tasnia T, Growe A (2025) A systematic literature review of water-sensitive urban design and flood risk management in contexts of strategic urban sustainability planning. Land 14(2):224. https://doi.org/10.3390/land14020224
31. UNICEF. (2017). Progress on drinking water. Sanitation and hygiene, 9–10.
32. Wong TH (2006) Water sensitive urban design-the journey thus far. Australas J Water Resour 10(3):213–222
33. Wong THF, Brown RR (2009) The water sensitive city: principles for practice. Water Sci Technol 60(3):673–682. https://doi.org/10.2166/wst.2009.436

34. World Bank (2018) Implementation completion and results report: Metro Colombo urban development project. World Bank, Washington, DC
35. World Health Organization (2017) World health statistics 2017: Monitoring health for the SDGs, Sustainable Dev Goals. Geneva. https://www.who.int/publications/i/item/9789241565486
36. Wwap U (2012) World water assessment programme: the United Nations world water development report 4: managing water under uncertainty and risk. UNESCO, Paris

Part II
Ecological Systems, Coastal Resilience, and Inter-Scalar Urban Futures

Small Coastal Cities Under Climate Pressure: Nature as Infrastructure and Lessons from Morondava, Avarua, Honiara, and Cairns

Ali Cheshmehzangi, Xin Wang, Umberto Baresi, Alessio Russo, Tian Tang, Amirhosein Shabani, Shima Taheri, and Alireza Mohammadi

1 Introduction: Small Coastal Cities at the Frontline of Climate Risk

Small coastal cities are among the urban systems most exposed to the impacts of climate change and natural hazards, yet they remain comparatively marginal within academic research and policy debates on urban resilience. In contrast, major coastal megacities like Shanghai, Jakarta, or New York have been the subject of most of the international literature on urban climate adaptation (Pelling and Blackburn, [10, 18, 44]). Nonetheless, a sizeable share of urban inhabitants in island regions and

A. Cheshmehzangi
School of Architecture, Design and Planning, The University of Queensland, Brisbane, Australia

A. Cheshmehzangi (✉) · X. Wang
College of Architecture and Urban Planning, Tongji University, Shanghai, China
e-mail: A.chesh@uq.edu.au

U. Baresi · A. Russo
School of Architecture and Built Environment, Faculty of Engineering, Queensland University of Technology, Brisbane, Australia

Centre for Environment and Society, Queensland University of Technology, Brisbane, Australia

T. Tang
Independant Researcher, Sichuan and Chongqing, China

A. Shabani
Department of Urban Planning, Na.C. Islamic Azad University, Najafabad, Iran

S. Taheri
Department of Architecture, Isfahan (Khorasgan) Branch, Islamic Azad University, Isfahan, Iran

A. Mohammadi
Department of Geography and Urban Planning, University of Mohaghegh Ardabili, Ardabili, Iran

A. Cheshmehzangi et al. (eds.), *Nature-based and Climate Responsive Design for Urban Systems*, Urban Sustainability, https://doi.org/10.1007/978-981-92-2070-0_7

rising coastal nations are found in smaller coastal communities, which frequently have populations of fewer than a few hundred thousand [28]. They are often more exposed to environmental problems, but they have far less institutional capacity and financial resources to mitigate these risks [42, 53].

Because of its economic benefits, which include access to maritime trade routes, fisheries, tourism prospects, and lush deltaic landscapes, coastal zones have historically drawn urban growth. However, coastal communities are increasingly at the forefront of the effects of climate change due to these same geographic features. Urban communities near the water are increasingly at risk from rising sea levels, stronger tropical cyclones, coastal erosion, and compound flooding occurrences. The Intergovernmental Panel on Climate Change (IPCC) estimates that, under intermediate emission scenarios, the global mean sea level will rise by 0.44 to 0.76 metres by the end of the twenty-first century, having risen by about 20 centimetres since 1900 [27, 33]. Even relatively modest increases in sea level can significantly amplify storm surges, accelerate shoreline retreat, and increase the frequency of coastal flooding events.

These changes produce several levels of risk for tiny coastal cities. Their housing systems, urban infrastructure, and economic activity are frequently concentrated in low-lying floodplains or small coastal strips. Development is often forced into environmentally vulnerable areas like wetlands, river deltas, and mangrove areas due to rapid population increase and the rise of informal settlements. Urban government institutions, especially the ones that would need it the most, frequently have difficulty financing large-scale protective infrastructure like seawalls, drainage systems, or coastal barriers, or successfully regulating land use [48]. Recent research shows that climate adaptation in coastal cities is often slow, limited in scope, and achieving insufficient outcomes, particularly in smaller urban centres that frequently lack the financial, technical, and institutional capacity to implement robust responses [16, 55]. For local climate adaptation to be effective, capacity is required at two key stages: the authority to act and the capacity to adapt, both depending on access to resources, professional networks, and supportive socio-political systems [46].

Tropical storms and cyclones pose a serious danger to urban resilience in many coastal places, especially in the Indian Ocean and Pacific Island systems. In a number of ocean basins, the intensity of tropical cyclones has increased along with the extreme rainfall events that are linked to these storms [30]. Housing, transportation infrastructure, energy systems, and local economy are frequently disrupted in small coastal cities situated in cyclone-prone areas. Strong winds and low air pressure can generate storm surges, which can flood cities, poison freshwater sources, and seriously harm coastal infrastructure. These occurrences are anticipated to result in more regular and damaging flooding in low-lying coastal areas when combined with sea-level rise [23, 36].

Even though conventional technical solutions like seawalls and hard coastal defences are still crucial parts of urban protection plans [11], they are sometimes expensive and challenging to put into practice in smaller communities with tighter budgets. Furthermore, stiff infrastructure can occasionally have unforeseen negative effects on the environment, including downstream erosion, habitat damage, and changed sediment flows. The potential of ecosystem-based adaptation approaches has been highlighted by environmental scientists and urban planners in response to these difficulties. These tactics acknowledge that natural coastal systems can serve as protective infrastructure, mitigating the effects of climate hazards while concurrently promoting local livelihoods and biodiversity [40, 52].

This growing awareness of coastal resilience is reflected in the idea of nature as infrastructure. Natural buffers that disperse wave energy, stabilize shorelines, and lessen the effects of storm surges include mangrove forests, coral reefs, tidal wetlands, and coastal vegetation systems. For example, mangroves can lower wave heights by up to 66% over very short coastal forest distances [40]. Similar to offshore breakwaters, coral reef systems can absorb up to 97% of wave energy before it reaches the shoreline [15]. These ecosystems offer social and economic advantages beyond their protective roles, such as increased fisheries productivity, carbon sequestration, and tourism potential. Therefore, in many coastal places, protecting and restoring these natural systems has grown in importance as part of climate adaptation initiatives.

This chapter examines how nature-based infrastructure can contribute to climate resilience in four small coastal cities in the Pacific and Indian Ocean regions: Cairns in northern Australia, Avarua in the Cook Islands, Honiara in the Solomon Islands, and Morondava in Madagascar. These cities have a number of things in common, despite the fact that their economic resources, political systems, and patterns of urban development vary greatly. Each is situated in a tropical coastal region that is vulnerable to sea-level rise, storm surges, and cyclones. For both commercial activity and environmental preservation, each greatly depends on the nearby coastal ecosystems. In addition, each city demonstrates a unique strategy for mitigating coastal risk through urban planning, ecosystem preservation, and adaptation tactics.

In the context of a developing nation, Morondava is a small coastal city where deltaic landscapes and mangrove ecosystems are essential for shielding urban areas from the effects of storms. The Cook Islands' administrative hub, Avarua, serves as an example of the difficulties confronted by small island capitals that rely heavily on coral reef systems and lagoon ecosystems for coastal protection but have little land availability. The challenges of urbanization and the spread of informal settlements in environmentally fragile coastal zones are reflected in Honiara, the capital of the Solomon Islands, which is expanding quickly. In contrast, Cairns is a more institutionally developed coastal city where efforts to adapt to climate change are aided by sophisticated planning processes and environmental protection techniques.

The chapter seeks to illustrate how ecological systems serve as essential elements of urban infrastructure in coastal areas by comparing these four cities. The analysis shows that while engineering solutions are still crucial, protecting and restoring natural ecosystems may offer some of the most practical and profitable ways to increase resilience in small coastal cities that are facing growing climate hazards.

2 Nature-Based Coastal Protection in Small Urban Systems

Historically, engineered infrastructures like levee systems, seawalls, groynes, and breakwaters have been linked to coastal protection. These buildings continue to be crucial parts of coastal defence, especially in heavily populated areas where people and valuable infrastructure need quick protection from flooding and storm surges. However, these solutions are frequently too costly, technically difficult, or physically impracticable for many small coastal communities, especially those in poor or island situations. This constraint to plan and implement adaptation is especially acute in small coastal towns and municipalities, which often face high climate exposure but are thwarted by limited institutional, financial, and technical capacity [16, 17]. Increased focus on ecosystem-based adaptation (EbA) and nature-based solutions (NbS) as workable and frequently more sustainable alternatives for coastal resilience [11, 31] has resulted from the growing awareness of these constraints.

The use, conservation, or restoration of natural ecosystems to lessen the effects of coastal risks while also providing social and environmental advantages is known as 'nature-based coastal protection' [7, 25]. Importantly, the literature increasingly emphasizes that nature-based methods should not be understood simply as substitutes for traditional infrastructure, but also as part of hybrid green-grey approaches, suitable in specific environmental conditions [38, 39]. These methods are based on the idea that natural mechanisms have developed over eons to control coastal activities, absorb shocks from the environment, and preserve ecological equilibrium. Nature-based methods aim to include ecological functions into urban resilience planning instead than substituting these systems with inflexible infrastructure [52]. Although the terminology of nature-based solutions is relatively recent, the practice of working with ecological processes as interconnected systems across multiple spatial scales has a long intellectual tradition in landscape architecture and planning, with applications in coastal protection recognized for almost 60 years [12, 34, 50, 59]. This systems-based perspective is most clearly articulated in Ian McHarg's Design with Nature, which emphasized that landscape performance and resilience depend on understanding and integrating biophysical processes from site to regional scales, particularly river-basin systems [34, 59]. Contemporary nature-based coastal protection can therefore be understood as an evolutionary reformulation of these principles, reframed through the ecosystem services framework to operationalize ecological processes as climate adaptation infrastructure [59]. Recent international guidance further reinforces this approach, highlighting that the effectiveness of nature-based coastal protection depends on integration across multiple spatial scales, especially at the river-basin level where upstream, urban, and coastal processes are intrinsically connected, and that the suitability and performance of NbS vary according to city typologies such as mountain, river, delta, and coastal settings [58].

A number of different kinds of coastal ecosystems are especially crucial for safeguarding urban shorelines. One of the most well-researched natural defences is mangrove forests (Fig. 1), which are prevalent in tropical and subtropical coastal areas. While their above-ground vegetation reduces storm surges and wave energy, their thick root systems stabilize sediments and lessen coastline erosion. Mangrove belts can considerably lessen wave height and storm impacts along coastal towns, reducing damage to dwellings and infrastructure during extreme weather events, according to empirical research [40]. Mangrove ecosystems contribute to coastal resilience through interacting ecological and geomorphological processes. Their complex root systems facilitate sediment trapping and vertical accretion, promoting shoreline stabilization and the maintenance of coastal geomorphology ([14]; Singh et al., [51]).

Mangroves also provide key ecosystem services. They function as critical nursery habitats that sustain fisheries productivity and contribute to water quality regulation through sediment retention and the removal or transformation of nutrients, heavy metals, and organic pollutants [14, 60]. The belowground compartment is central to ecosystem functioning, as microbial communities within mangrove sediments mediate key biogeochemical processes, including organic matter mineralization, nutrient cycling, and contaminant transformation [14]. Mangroves are also recognized as major blue-carbon ecosystems, as their total ecosystem carbon stocks often exceed those of many terrestrial forests on a per-unit-area basis, positioning them as nature-based solutions that integrate coastal protection, ecosystem services, and

Fig. 1 Mangrove vegetation at the northern tip of Bribie Island, Queensland, Australia, facing Golden Beach in Caloundra. The photograph captures mangrove shrubs established on intertidal mudflats with visible pneumatophores during low tide, characteristic of sheltered coastal and estuarine systems. *Source* Photo taken by Alessio Russo

climate change mitigation and adaptation [14]. However, recent global assessments highlight that mangroves and the ecosystem services they deliver, including coastal protection, are increasingly exposed to climate-related risks arising from intensified tropical cyclone disturbance regimes and accelerating sea-level rise [26].

In many tropical island ecosystems, coral reef systems serve a complementary defensive function [5]. Reefs, which are found offshore, serve as natural breakwaters by dispersing wave energy before it reaches coastal regions. According to research, coral reefs can absorb up to 97% of wave energy, lowering the danger of erosion and coastal flooding [15]. Maintaining healthy reef ecosystems becomes essential to long-term coastal resilience for small island settlements, where land is scarce and infrastructural investments may be limited.

Coastal marshes and tidal wetlands also play a major role in protecting the coast. These ecosystems absorb storm energy across shallow coastal plains, accumulate elevation over time, and trap sediments. By promoting biodiversity, filtering pollutants, and sustaining regional livelihoods through fishing and aquaculture, wetlands offer additional ecological services. However, urban growth and land reclamation have led to the loss or degradation of these natural systems in many coastal areas, making coastal towns more susceptible to storm damage and floods [4].

Including such ecosystems in urban planning has several benefits for tiny urban systems. When compared to constructed infrastructure, nature-based methods are frequently more flexible and self-maintaining. Ecosystems can offer long-term protection without the need for continuous structural reinforcement because they can expand, regenerate, and change in response to shifting environmental conditions [2, 6, 32, 37]. Additionally, these strategies are usually less expensive to build and maintain, which makes them especially appropriate for cities with tight budgets depending on site-specific ecological conditions, design quality, governance arrangements, and the time available for ecosystems to mature.

Nature-based coastal preservation relies heavily on local ecological knowledge and community stewardship in addition to their ecological and engineering roles. Local communities in many coastal societies have generations of expertise about sustainable resource use, ecosystem management, and shoreline dynamics. Local involvement can greatly increase the efficacy and durability of ecosystem-based adaptation techniques, as shown by community-based mangrove restoration projects, reef protection efforts, and coastal vegetation initiatives [49]. Coastal ecosystems are more likely to be preserved and eventually restored when communities actively participate in their preservation. Because nature-based interventions do not automatically produce inclusive outcomes, poorly designed participation processes can hide and fuel power inequalities, exclude marginalized groups, or unevenly distribute benefits and burdens [43, 45]. These risks are particularly relevant in small coastal cities, where adaptation decisions can directly affect land access, livelihoods, and the rights of vulnerable residents [43, 57].

However, there are drawbacks to using natural remedies. Degraded systems may take years or even decades to recover, and ecosystem restoration may take a long time to reach full protective capacity. In order to stop further deterioration of coastal ecosystems, effective implementation also necessitates coordinated governance frameworks [47], land-use management, and environmental monitoring. Because of this, nature-based infrastructure works best when combined with more comprehensive planning and policy frameworks that promote long-term ecological management.

To compare the four case study cities, this chapter uses the idea of nature as infrastructure as its analytical framework. The chapter addresses ecosystems as useful elements of urban resilience systems rather than only as environmental resources. Thus, in coastal urban environments, mangroves, coral reefs, wetlands, and coastal vegetation are viewed as infrastructure networks that serve protective, ecological, and socioeconomic functions. At the same time, recent critical scholarship reminds that framing nature as infrastructure is not politically neutral, because it can also instrumentalize ecological systems in the service of development or resilience agendas [41] In order to mitigate climate threats, the comparative analysis looks at how Morondava, Avarua, Honiara, and Cairns interact with their local ecosystems. The chapter aims to illustrate the advantages and disadvantages of ecosystem-driven coastal resilience methods in small urban systems by examining how various governance capacities, financial resources, and environmental conditions influence the application of nature-based protection.

3 Case Study Contexts: Four Coastal Cities across the Indian and Pacific Oceans

This section introduces the four case studies, outlining their geographic setting, urban development patterns, and climate vulnerabilities. While all cities share susceptibility to coastal risks, they reflect distinct socioeconomic and governmental contexts. The intricate relationship between urban growth, environmental systems, and climate vulnerability is exemplified by small coastal cities in the Pacific and Indian oceans. Despite the fact that these cities differ greatly in terms of their governance systems, economic capacity, and urban planning frameworks, they all have one thing in common: being close to the water exposes them to growing climate hazards. Despite representing different socioeconomic contexts, the four case studies discussed in this chapter, Morondava in Madagascar, Avarua in the Cook Islands, Honiara in the Solomon Islands, and Cairns in Australia, show similar patterns of exposure to coastal hazards and reliance on nearby ecosystems for protection (See Fig. 2).

Situated near the Mozambique Channel on Madagascar's western coast, Morondava is a small coastal city surrounded by mangroves and a delta. The city is close to the mouth of the Morondava River, where the coastal environment is shaped by mangrove forests, estuarine systems, and tidal flats. Because of its close vicinity to the world-renowned Avenue of the Baobabs, Morondava's economy is mostly dependent on fishing, small-scale agriculture, and tourism. However, the city is becoming more susceptible to flooding, coastal erosion, and tropical cyclones. It is especially vulnerable to climate-related risks due to its comparatively small urban infrastructure

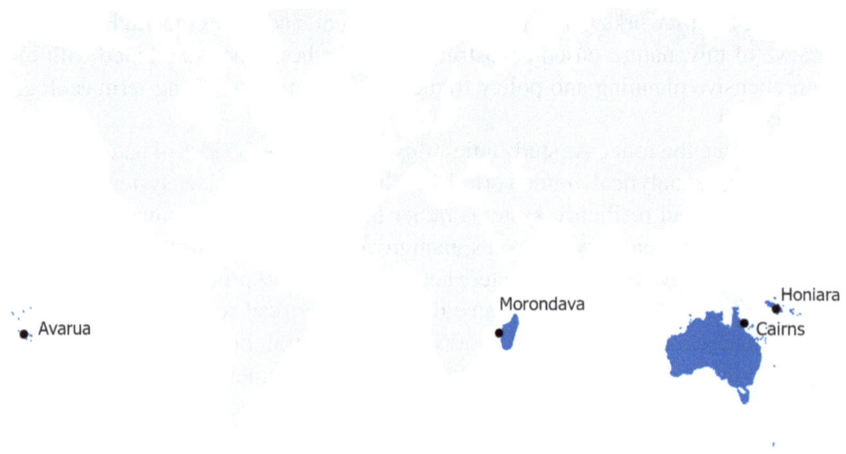

Fig. 2 Location of the four case studies

and planning capabilities. The city's surrounding mangrove forests are crucial for maintaining shorelines and shielding coastal communities from storm surges.

Avarua, the capital of the Cook Islands, is located on the northern coast of the island of Rarotonga. In contrast to Morondava's deltaic scenery, Avarua is located in a volcanic island environment with narrow coastal plains surrounded by steep interior mountains. A large portion of the island is surrounded by lagoon ecosystems and coral reef systems, which act as natural barriers to disperse wave energy before it reaches the shoreline. The Cook Islands' administrative, economic, and transportation centre is Avarua, where tourism is the main industry. Large-scale infrastructure development is hampered by the island's tiny land area and remote location, despite its comparatively good environmental management and community governance. Because of this, Avarua is still quite vulnerable to storm surge occurrences brought on by tropical cyclones, coastal erosion, and sea-level rise.

Honiara, the capital city of the Solomon Islands, presents a different urban dynamic. Located on Guadalcanal Island's northern shore, Honiara has seen a sharp increase in its urban population in recent years. A small coastal corridor between the sea and hilly terrain is where much of the city's growth is located. The growth of informal settlements in environmentally fragile locations, such as coastal lowlands and riverbanks that are prone to flooding, has been facilitated by this geographical constraint. Tropical cyclones, heavy rainfall, flooding, and landslides are just a few of the climate-related hazards that Honiara must deal with. Important ecosystem services are provided by the mangrove forests and coastal wetlands that surround the city, but these natural systems are under growing stress due to urban growth and environmental degradation. Governance and infrastructure capacity remain limited relative to the scale of urban growth, further complicating climate adaptation efforts.

Situated in Australia's tropical northern Queensland, Cairns is one of the region's more institutionally advanced coastal cities. Cairns is located near the Great Barrier

Reef and bordered by vast mangrove wetlands along Trinity Inlet and other coastal floodplains. The city serves northern Australia's tourism, transportation, and service sectors as a significant regional economic centre. Cairns is nevertheless very vulnerable to tropical cyclones, heavy rains, and coastal flooding despite having comparatively robust infrastructure and administration. The region's infrastructure and urban systems continue to be vulnerable, as seen by recent storm events and major flooding bouts. Cairns, however, enjoys the advantages of sophisticated coastal management techniques, environmental monitoring systems, and planning frameworks for climate adaptation when contrasted to many smaller island communities.

Taken together, these four cities show a variety of coastal urban environments seen throughout the Pacific and Indian oceans. They are exposed to similar climate threats and rely largely on adjacent ecosystems for coastal protection but have quite different economic capacities and governance arrangements. Comparing these cities offers important insights into how various urban systems interact with natural coastal environments and how ecosystem-based strategies could promote resilience in a variety of coastal contexts (See Table 1).

3.1 Morondava, Madagascar

The coastal city of Morondava is situated along the Mozambique Channel on Madagascar's western coast. The city is located close to the mouth of the Morondava River, where tidal flats, estuaries, and vast mangrove habitats have been sculpted by deltaic processes. Morondava, which is home to between 40,000 and 60,000 people, serves as the Menabe region's centre for local trade, agriculture, and fishing. Important mangrove forests that assist local communities economically and ecologically are supported by the surrounding coastal environment [29]. Mangroves stabilise sediments along the shore, boost fisheries output, and offer natural defence against storm surges and coastal erosion. As an example, [29] study provides detailed mapping and analysis of Madagascar's mangrove systems, including their ecological dynamics and pressures such as land conversion and resource extraction.

Many Morondava inhabitants still rely heavily on coastal ecosystems for their livelihoods. Mangrove resources are frequently utilized for fuelwood, building materials, and small-scale aquaculture operations, while artisanal fishing is a significant source of income and food security. The neighbouring natural features, especially the internationally renowned Avenue of the Baobabs, which has grown to be one of Madagascar's most popular landscapes, have also contributed to the growth of tourism [35]. Despite these economic endeavours, Morondava is still a comparatively small metropolitan system with little infrastructure and resources for managing urban planning.

The city is facing more challenges due to climate exposure. Tropical cyclones that originate in the Indian Ocean and move across the Mozambique Channel frequently affect Western Madagascar. Particularly in low-lying deltaic regions like Morondava,

Table 1 Comparative overview of coastal context, climate exposure, and governance capacity in Morondava, Avarua, Honiara, and Cairns

Dimension	Morondava (Madagascar)	Avarua (Cook Islands)	Honiara (Solomon Islands)	Cairns (Australia)
Geographic setting	Coastal delta on the Mozambique Channel; located near mangrove estuaries and river systems	Small coastal capital on the island of Rarotonga with volcanic terrain, lagoon, and coral reef protection	Coastal city on Guadalcanal Island along a narrow coastal strip between mountains and sea	Tropical coastal city in Northeast Queensland located around trinity inlet and adjacent coastal floodplains
Urban scale	Small regional city (approx. 40,000–60,000 residents)	Very small national capital (approx. 10,000–15,000 in the urban core)	Medium-sized Pacific capital (approx. 80,000–100,000 metropolitan population)	Regional city (approx. 150,000 residents in greater Cairns area)
Economic base	Fisheries, small-scale agriculture, tourism (notably the avenue of the baobabs), and local trade	Tourism-led economy, public administration, small-scale fisheries, and services	Government administration, port and logistics activities, fisheries, trade, and significant informal economy	Tourism (especially reef tourism), services, port logistics, aviation gateway, and regional economic services
Coastal ecosystem context	Extensive mangrove systems, tidal flats, river delta wetlands, and sedimentary coastal landscapes	Coral reef systems, lagoon environments, narrow coastal plains, and tropical marine ecosystems	Mangrove forests, coral reefs, coastal wetlands, and riverine estuaries	Large mangrove wetlands, estuarine ecosystems, and proximity to the great barrier reef
Primary climate risks	Tropical cyclones, coastal erosion, flooding, sea-level rise, and storm surge	Tropical cyclones, storm surge, coastal erosion, sea-level rise, and coastal inundation	Tropical cyclones, intense rainfall, flooding, landslides, and coastal erosion	Tropical cyclones, extreme rainfall events, storm surge, coastal flooding, and riverine flooding

(continued)

Table 1 (continued)

Dimension	Morondava (Madagascar)	Avarua (Cook Islands)	Honiara (Solomon Islands)	Cairns (Australia)
Urban expansion pattern	Informal coastal expansion and migration from rural hinterlands	Limited urban expansion due to island geography and restricted land availability	Rapid expansion of informal settlements along coastal corridors and hillsides	Planned suburban expansion supported by zoning regulations and urban growth management
Nature-based protection systems	Mangrove ecosystems and coastal vegetation acting as buffers against waves and storm surges	Coral reefs and lagoon systems functioning as natural breakwaters that dissipate wave energy	Mangroves and coral reef systems providing shoreline stabilization and ecosystem services	Mangrove wetlands, coastal vegetation buffers, and ecosystem conservation within coastal management programmes
Infrastructure capacity	Limited urban infrastructure, limited flood protection, and constrained resources for coastal defence	Moderate infrastructure capacity but constrained by geographic isolation and limited land	Developing infrastructure systems with capacity challenges and increasing urban pressure	Advanced infrastructure systems, strong disaster preparedness frameworks, and developed coastal engineering
Governance capacity	Developing governance systems and evolving urban planning frameworks	Small-island governance structures with strong community networks and traditional decision-making systems	Limited institutional capacity combined with rapid urbanization pressures	Strong planning institutions, integrated coastal management frameworks, and established climate adaptation policies

(continued)

Table 1 (continued)

Dimension	Morondava (Madagascar)	Avarua (Cook Islands)	Honiara (Solomon Islands)	Cairns (Australia)
Adaptation strategies	Mangrove restoration programmes, ecosystem-based coastal protection, and community-led adaptation	Coastal setback policies, coral reef conservation initiatives, and coastal planning controls	Mangrove restoration initiatives and community-based climate adaptation strategies	Integrated coastal zone management, flood mitigation infrastructure, wetland conservation, and resilience planning
Role in regional urban system	Regional coastal hub supporting fisheries and trade in Western Madagascar	National administrative and economic Centre of the Cook Islands	Primary political, economic, and administrative Centre of the Solomon Islands	Major regional economic hub and gateway city for northern Australia

these storms can cause severe rainfall, storm surges, and coastal flooding. Additionally, coastal erosion has been noted along several shoreline segments, in part due to shifting sediment flows and wave dynamics. Additionally, in certain places, natural coastal protection has been compromised by the destruction of mangrove ecosystems brought on by resource exploitation and land conversion. Mangrove system restoration and preservation have grown in importance as part of the region's climate resilience plans as sea levels rise and extreme weather events become more frequent.

3.2 *Avarua, Cook Islands*

Situated on the northern shore of Rarotonga, the largest island in the country, Avarua is the administrative, political, and economic hub of the Cook Islands. Avarua is situated on a small coastal plain inside a volcanic island setting, in contrast to many coastal cities that grow on vast continental landmasses. There is little flat area suitable for urban development because the island's interior is mountainous and climbs abruptly from the coast. Because of this, the majority of the island's houses, infrastructure, and economic activity are concentrated along a small ring of coastline land, or known as 'coastal ring settlement' [15].

The existence of coral reef systems and lagoon waters that ring a large portion of Rarotonga is a defining feature of Avarua's coastal environment. By absorbing wave energy and lessening the effects of storm surges and ocean waves, these reefs are crucial to the shoreline's protection. Additionally, the lagoon ecology sustains fisheries and marine biodiversity, both of which support local livelihoods [19]. Since tourism is the Cook Islands' main industry, the stability of the island's economy is directly related to the state of these reef and lagoon systems.

Even with coral reefs' natural defence, Avarua is nevertheless quite susceptible to climate-related risks. The Cook Islands are part of the South Pacific cyclone belt, and the island is occasionally hit by tropical cyclones that bring with them powerful winds, a lot of rain, and coastal flooding. Because a large portion of the island's infrastructure, including roads, public buildings, and tourism infrastructure, is situated close to the coast, sea-level rise poses a longer-term threat. Relocating development inland is further hampered by the scarcity of land.

On Rarotonga, community governance and environmental management are crucial to preserving coastal resiliency. To preserve reef ecosystems and lessen shoreline damage, local authorities have implemented marine conservation programmes and coastal setback regulations. Ecological systems that support local livelihoods and coastal protection have also been sustained through community stewardship of marine resources, including traditional lagoon management techniques.

3.3 Honiara, Solomon Islands

Situated on Guadalcanal Island's northern shore, Honiara serves as the main political, economic, and transportation hub of the Solomon Islands. The city has had consistent population growth since attaining independence in 1978, fuelled by migration from rural to urban areas and growing economic prospects. With an estimated population of more than 80,000 people, Honiara is currently one of the Pacific region's fastest-growing metropolitan centres [22].

Urban growth patterns are heavily influenced by the physical topography of the city. Honiara is located between rugged mountains and the sea in a small coastal strip [1]. Due to the scarcity of land, informal settlements have grown around riverbanks, hillside regions, and coastal plains. Inadequate infrastructure, including drainage systems, sanitary facilities, and sturdy dwelling constructions, is present in many of these villages. Large portions of the metropolitan population are therefore more susceptible to environmental risks.

In Honiara, climate-related concerns are especially serious. The Solomon Islands are situated in a part of the South Pacific that is prone to cyclones, and heavy rains often result in landslides and flooding. Low-lying towns and urban infrastructure close to the coast are also at risk from storm surges and coastal floods. The surrounding habitats, including mangrove forests and coastal wetlands, which historically offered natural protection against erosion and storm impacts, are under growing pressure due to the city's rapid expansion.

The size of these habitats around Honiara has decreased due to resource extraction and urbanization, raising concerns about mangrove degradation in recent decades. A number of community-based conservation and restoration projects have surfaced recently in recognition of the mangroves' defensive role. These initiatives seek to improve local understanding of the role ecosystems play in mitigating climate risks while restoring damaged coastal habitats. However, the effectiveness of such efforts remains closely tied to broader governance challenges, including land management, infrastructure investment, and urban planning capacity.

3.4 Cairns, Australia

A significant commercial entry point for northern Australia and the Great Barrier Reef region, Cairns is a regional coastal city in tropical northern Queensland, Australia. The settlement is located near Trinity Inlet, a tidal estuary encircled by low-lying floodplains and vast mangrove wetlands. With a population of over 150,000, Cairns is a major hub for the local transportation, tourism, and service sectors [3].

One of the things that makes Cairns unique is its natural surroundings. The city is situated between two globally important natural systems: the Wet Tropics rainforest inland and the Great Barrier Reef offshore. The tidal wetlands and mangrove forests that round Trinity Inlet serve as a crucial natural buffer that shields the shoreline from storm damage and erosion. In the coastal environment, these ecosystems also sustain biodiversity, fisheries, and the control of water quality [13].

Cairns is nevertheless very vulnerable to climatic threats, even with its comparatively robust infrastructure and governance structures. Tropical cyclones and periods of heavy rainfall that can cause extensive floods are frequent occurrences in Northern Queensland. Because of the city's floodplain location, drainage and stormwater management are essential elements of urban development [13]. Tropical cyclone-related severe floods have occasionally caused disruptions to the region's housing, transportation, and infrastructure.

Compared to many smaller coastal communities in the Pacific, Cairns has a more sophisticated institutional framework for climate adaptation [9]. Integrated coastal zone management policies that incorporate land-use planning, disaster risk reduction, and environmental conservation have been put into place by local and state authorities through instruments such as Our Cairns Coast: Adapting for the Future, the Cairns Climate Change Strategy 2030, CairnsPlan 2016, and Queensland's coastal hazard planning guidance [8, 9]. These initiatives include coastal setback laws, wetland conservation initiatives, and mangrove protection rules. Disaster preparedness is also aided by coastal hazard mapping, adaptation pathways, and environmental monitoring embedded in these planning processes. Cairns thus serves as an illustration of how urban development and ecological preservation might be combined into a more all-encompassing framework for coastal resilience.

4 Comparative Analysis: Nature as Infrastructure in Coastal Adaptation

Natural ecosystems in coastal regions carry out tasks that are often performed by manmade infrastructure. Wetlands, mangrove forests, coral reefs, and coastal vegetation systems stabilize shorelines, control sediment flows, lessen the effects of storm surges, and attenuate wave energy. Rather than being peripheral environmental assets, these ecosystems become essential parts of urban resilience measures when they are viewed as nature-based infrastructure. The four case studies, Morondava, Avarua, Honiara, and Cairns, show how natural systems can function as protective infrastructure in various governmental and socioeconomic circumstances.

The kind of dominant environment that protects the coast is a crucial point of comparison between the cities. Mangrove ecosystems serve as the main line of defence in Morondava and Honiara. In deltaic and estuary settings, mangrove root systems stabilize coasts, trap sediments, and lessen wave energy. The primary natural defence in Avarua is provided by coral reef systems and lagoon habitats, which dissipate offshore wave energy before it reaches the shoreline. Cairns is an example of a hybrid system that protects metropolitan areas by combining regulated coastal planning frameworks, mangrove wetlands, and estuary ecosystems.

Environmental management techniques and governance capabilities have a major role in how well these natural systems function as infrastructure. Mangrove ecosystems in Morondava are still essential for protecting the coast, but resource extraction and land conversion are putting more and more strain on them. Investment in coastal infrastructure or extensive restoration is hampered by a lack of planning capacity and municipal resources. Therefore, preserving the ecological integrity of nearby mangrove systems is crucial to coastal resiliency.

Similar ecological conditions are found in Honiara, although there is more urban pressure there. Mangrove areas and coastal vegetation have been cleared in several regions of the city due to rapid population growth and the rise of informal settlements. Natural coastal defences have been reduced by the destruction of these ecosystems, making them more susceptible to storm surge and flooding. Stronger institutional coordination and better land management are necessary for the success of recent community-based restoration projects, which seek to buck these trends.

Avarua, on the other hand, serves as an example of how natural marine ecosystems can serve as offshore defence infrastructure. Rarotonga's coral reefs serve as natural breakwaters, absorbing incoming wave energy to lessen the effects of storm surges and shoreline erosion. In order to sustain these protective systems, community-based reef management and marine conservation regulations have become crucial. However, because coral reefs are extremely vulnerable to physical storm damage, ocean warming, and acidification, long-term resilience depends on more comprehensive climate mitigation initiatives and ecosystem management.

Cairns serves as an example of how a more thorough planning framework might incorporate nature-based infrastructure. Trinity Inlet's mangrove wetlands are safeguarded by land-use laws, conservation initiatives, and coastal management plans.

These ecosystems sustain fisheries and biodiversity while lessening the effects of storm surges and safeguarding urban shorelines. The city benefits from sophisticated drainage systems, flood mitigation infrastructure, and hazard mapping in addition to ecological protection. This hybrid adaptation concept, in which natural ecosystems supplement rather than replace constructed defences, is exemplified by the mix of ecological systems and artificial infrastructure.

Local stewardship and community involvement are another crucial aspect of comparison. Local communities are essential to the conservation and regeneration of mangroves in Morondava and Honiara. Coastal ecosystems are maintained by community monitoring programmes, small-scale fisheries management, and traditional ecological knowledge. Local management of lagoon and reef ecosystems in Avarua is still intimately linked to sustainable tourism and marine conservation initiatives. Even though Cairns has a more formal government system, community participation in conservation and environmental monitoring initiatives is beneficial.

In general, the comparison identifies three different ways that nature serves as infrastructure for coastal adaptation. First, ecosystem reliance, in which cities are directly dependent on natural systems because they lack the resources or expertise to build artificial defences. The second is ecosystem stewardship, in which ecological protection systems are maintained by community-based governance. Third, ecological systems are combined with formal planning frameworks and technical solutions in hybrid infrastructure models.

These paths show that whereas ecological systems offer substantial protective advantages, governance capability, environmental management, and urban growth constraints significantly influence their efficacy. Therefore, in addition to ecological studies, institutional and socioeconomic settings must be taken into account in order to comprehend how natural ecosystems serve as infrastructure (Table 2).

5 Discussion: Governance, Capacity, and the Future of Coastal Urban Resilience

The comparison of Morondava, Avarua, Honiara, and Cairns reveals a crucial finding for modern coastal planning: the efficiency of nature-based infrastructure is strongly correlated with institutional coordination, financial resources, governance capacity, and ecological conditions. Although mangroves, coral reefs, wetlands, and coastal vegetation offer significant natural defences against climate threats, their protective roles can only be maintained with long-term policy support and efficient management frameworks. Therefore, rather than being a merely natural phenomenon, coastal resilience must be viewed as the outcome of interactions between ecological systems and governance structures. This broader pattern is consistent with recent scholarship showing that local climate adaptation depends not only on ecological assets, but also on the authority to act, institutional capacity, and access to financial and technical resources [16, 46].

Table 2 Nature as infrastructure in coastal adaptation: comparative overview of four cities

City	Primary natural infrastructure	Key protective functions	Governance and management approach	Major challenges
Morondava	Mangrove forests and estuarine wetlands	Wave attenuation, shoreline stabilization, sediment trapping	Limited institutional capacity; emerging ecosystem restoration efforts	Mangrove degradation, cyclone exposure, limited infrastructure investment
Avarua	Coral reef systems and lagoon environments	Offshore wave energy dissipation, shoreline protection	Marine conservation policies and community-based reef stewardship	Coral bleaching, sea-level rise, limited land for relocation
Honiara	Mangrove ecosystems and coastal wetlands	Flood buffering, shoreline stabilization, ecosystem services	Community-based mangrove restoration with limited formal planning capacity	Rapid urban growth, informal settlement expansion, ecosystem degradation
Cairns	Mangrove wetlands and estuarine systems integrated with planning	Storm surge reduction, flood regulation, biodiversity support	Integrated coastal zone management and environmental planning	Increasing cyclone intensity, floodplain exposure, climate variability

The four cities' institutional ability to incorporate ecological systems into official planning frameworks is one of their most notable differences. Cairns exhibits a comparatively sophisticated governance framework where environmental legislation and coastal planning incorporate ecosystem conservation. A coordinated approach including local government, state agencies, and scientific monitoring programmes includes integrated coastal zone management rules, floodplain zoning, and mangrove conservation. With the help of financial investment and planning legislation, this governance capacity enables natural ecosystems to be officially acknowledged as elements of urban infrastructure. This institutional capacity is reflected in formal planning and adaptation instruments such as Our Cairns Coast: Adapting for the Future, the Cairns Climate Change Strategy 2030, CairnsPlan 2016, and Queensland's coastal hazard planning guidance, all of which link hazard management with ecosystem protection and land-use planning [8, 9]. Earlier academic work on Trinity Inlet likewise shows that coordinated mangrove protection and coastal planning have long been understood as central to both ecological protection and urban resilience in Cairns [56].

Morondava and Honiara, on the other hand, work under governance settings with inadequate institutional resources and planning enforcement. Mangrove ecosystems in Morondava offer crucial defence against storm surges and coastal erosion, but their ability to be managed is hampered by a lack of municipal finance, disjointed land governance, and conflicting livelihood demands. In a similar vein, Honiara's mangrove habitats, which traditionally served as natural coastal buffers, are under tremendous strain due to the city's fast urban population increase and the spread of informal settlements. Long-term ecosystem protection is nevertheless threatened by inconsistent land-use legislation and enforcement, notwithstanding the recent emergence of community-based restoration initiatives. These conditions closely reflect the wider literature on smaller municipalities, which shows that high exposure to climate risk is often matched by weak implementation capacity, fragmented institutions, and difficulty mainstreaming adaptation into everyday planning practice [16, 55]. In Honiara specifically, recent urban resilience planning has tried to respond to these pressures through the Honiara Urban Resilience and Climate Action Plan and the Climate Resilient Honiara programme, but these initiatives also underline how dependent long-term ecosystem protection remains on sustained institutional support and external partnerships [54].

Avarua occupies an intermediate position within this governance spectrum. The city benefits from comparatively robust community governance institutions and environmental stewardship practices as the Cook Islands' administrative hub. A proactive approach to preserving natural coastal defences is shown by marine conservation zones, lagoon protection projects, and reef management activities. However, large-scale infrastructure adaptation or urban relocation solutions are constrained by the island's limited land supply and economic reliance on tourism. In this situation, coral reef ecosystems effectively function as offshore infrastructure, offering coastal protection and wave attenuation that would otherwise necessitate costly artificial alternatives. This position is reinforced by the Cook Islands' broader policy emphasis on integrated climate, marine, and disaster governance, including instruments such as Marae Moana and national climate adaptation planning frameworks [20, 21]. The continued importance of customary practices such as ra'ui also suggests that Avarua's resilience is shaped not only by state planning, but by the interaction between formal governance and longer-standing forms of community environmental stewardship [24].

Adaptation routes are also significantly impacted by financial resources. Hybrid techniques, which integrate ecological protection with engineered systems like enhanced drainage networks, stormwater management infrastructure, and coastal monitoring systems, can be implemented by cities with easier access to funds. Mangrove wetlands and estuarine ecosystems coexist with flood mitigation infrastructure and hazard mapping initiatives in Cairns, a prime example of this hybrid approach. On the other hand, because of their limited financial resources for extensive engineering interventions, Morondava and Honiara rely more on ecosystem-based protection. This supports the growing argument in the adaptation literature that nature-based measures are often most effective not as stand-alone substitutes for engineering, but as part of hybrid green-grey systems adapted to local environmental and institutional conditions [38, 39]. Recent comparative work on coastal cities similarly suggests that adaptation tends to remain slower and narrower in scope where resources for large-scale public infrastructural responses are limited [55].

Coastal resilience is also found to be significantly influenced by local knowledge and community involvement. Through fishing, resource harvesting, and cultural customs, local residents in many tiny coastal cities preserve intimate linkages with the surrounding ecosystems. Local stewardship can promote ecosystem recovery and improve coastal protection, as demonstrated by community-led mangrove restoration projects in Madagascar and the Solomon Islands. In a similar vein, the Cook Islands' traditional marine management techniques support the preservation of lagoon ecosystems and reef health. The efficacy of nature-based adaptation solutions can be increased by incorporating such local knowledge systems into formal planning frameworks.

Looking forward, increased cooperation at the municipal, national, and regional levels will be necessary to improve coastal resilience in small urban systems. In vulnerable coastal areas, ecological restoration, coastal monitoring, and urban resilience planning are increasingly supported by international climate adaptation programmes, development organizations, and scientific research networks. For cities dealing with complicated climate threats, these partnerships can offer financial support, technical know-how, and governance advice.

Future adaptation plans, however, are probably going to depend more on hybrid infrastructure solutions, in which constructed systems and natural ecosystems operate together as integrated resilience networks. Wetland conservation may improve urban flood management systems, mangrove restoration may supplement seawalls, and coral reef protection may lessen the need for manmade breakwaters. Effective coastal adaptation will necessitate creating urban systems that incorporate both biological and engineered solutions rather than seeing them as mutually exclusive.

In the conclusion, Morondava, Avarua, Honiara, and Cairns' experiences show that coordinating natural systems, government ability, and community involvement are critical to the future of coastal urban resilience. Although small coastal communities may be disproportionately vulnerable to climate change, they also have special chances to create adaptive plans that make use of natural infrastructure, which is a key element of urban resilience (Table 3).

6 Conclusions and Lessons for Small Coastal Cities

A fundamental truth of coastal urban development in the twenty-first century is demonstrated by the comparison of Morondava, Avarua, Honiara, and Cairns: natural ecosystems are not just environmental assets but also crucial elements of metropolitan infrastructure. By lowering wave energy, maintaining shorelines, controlling sediment flows, and buffering coastal flooding, mangrove forests, coral reefs, wetlands, and estuary systems serve protective purposes like those of manmade barriers. These biological systems generally serve as the first and best line of defence against climate threats in small coastal cities with limited institutional and financial resources.

The substantial correlation between governance capability and the efficacy of nature-based infrastructure is one of the main conclusions drawn from this study.

Table 3 Governance capacity and adaptation pathways in four coastal cities

City	Governance capacity	Financial resources	Role of Natural infrastructure	Adaptation approach	Key constraints
Morondava	Limited institutional planning capacity	Low	Mangrove ecosystems provide primary coastal protection	Ecosystem-dependent adaptation with emerging restoration initiatives	Resource constraints, mangrove degradation, limited urban planning enforcement
Avarua	Moderate governance supported by strong community institutions	Moderate	Coral reefs and lagoon systems act as offshore protective infrastructure	Marine conservation and ecosystem stewardship	Limited land availability, reliance on tourism economy
Honiara	Developing governance with urban growth pressures	Low to moderate	Mangrove wetlands historically protected coastal areas	Community-based restoration and emerging adaptation programmes	Informal settlement expansion, ecosystem degradation
Cairns	Strong planning institutions and integrated coastal management	High	Mangrove wetlands integrated into formal environmental planning	Hybrid infrastructure combining ecological and engineered systems	Increasing climate hazards and floodplain exposure

Cities like Cairns are better positioned to formally acknowledge and safeguard natural ecosystems as a component of their infrastructure systems because they have more robust planning institutions and integrated environmental regulations. Ecological assets can be protected and incorporated into long-term resilience plans through land-use management, coastal zoning, and environmental monitoring. On the other hand, cities with weak governance capabilities, like Morondava and Honiara, frequently rely significantly on ecosystem services without having enough financial or regulatory resources to adequately safeguard them. In these situations, urban susceptibility to the effects of climate change can be greatly increased by environmental deterioration.

A second important lesson concerns the role of community stewardship and local ecological knowledge in sustaining natural coastal defences. Through fishing, resource harvesting, and cultural customs, communities in many tiny coastal cities preserve intimate linkages with the surrounding ecosystems. Maintaining the health and resilience of ecosystems has been shown to depend on local involvement in mangrove restoration, reef conservation, and coastal vegetation protection. These experiences show that effective ecosystem-based adaptation necessitates genuine involvement with local communities whose livelihoods depend on these habitats in addition to scientific understanding and planning expertise.

Additionally, the comparison emphasizes the significance of hybrid adaptation systems that integrate engineering and ecological solutions. Even though nature-based infrastructure has many protective advantages, it isn't always enough to deal with all of the climatic risks that coastal communities face. The most resilient results are probably going to come from integrated strategies that incorporate wetland restoration, coral reef protection, mangrove conservation, and engineering infrastructure like flood barriers and drainage systems. Cairns is an illustration of how coordinated planning frameworks might incorporate such hybrid systems.

One of the most affordable and sustainable adaptation strategies for many small coastal cities in the Global South and Pacific regions is the preservation and restoration of natural ecosystems. However, achieving this requires stronger institutional coordination, long-term environmental management, and continued investment in community-based stewardship initiatives.

Lastly, experiences from Morondava, Avarua, Honiara, and Cairns indicate that a change in viewpoint will be necessary for the future of coastal urbanization. Cities must acknowledge ecosystems as vital infrastructural systems that support long-term urban resilience rather than viewing them as incidental environmental issues. Adopting this strategy presents a chance to match urban growth with ecological processes, resulting in coastal cities that are more socially inclusive, environmentally sustainable, and climate change adaptable. Within this framing, future research should focus on integrating nature-based infrastructure with early-warning systems, climate change modelling, long-term ecosystem monitoring, and dedicated funding mechanisms to support anticipatory planning and adaptive governance in small coastal cities, particularly as recent research indicates that climate-driven shifts in cyclone intensity and sea-level rise [26] are undermining the long-term stability and service-provision capacity of mangrove ecosystems.

References

1. Albert S, Leon JX, Grinham AR, Church JA, Gibbes BR, Woodroffe CD (2016) Interactions between sea-level rise and wave exposure on reef island dynamics in the Solomon Islands. Environ Res Lett 11(5):054011. https://doi.org/10.1088/1748-9326/11/5/054011
2. Alexander S, Aronson J, Whaley O, Lamb D (2016) The relationship between ecological restoration and the ecosystem services concept. Ecol Soc 21(1)
3. Alongi DM (2008) Mangrove forests: resilience, protection from tsunamis, and responses to global climate change. Estuar, Coast Shelf Sci 76(1):1–13. https://doi.org/10.1016/j.ecss.2007.08.024
4. Barbier EB, Hacker SD, Kennedy C, Koch E, Stier A, Silliman B (2011) The value of estuarine and coastal ecosystem services. Ecol Monogr 81(2):169–193
5. Beck MW, Losada IJ, Menéndez P, Reguero BG, Díaz-Simal P, Fernández F (2018) The global flood protection savings provided by coral reefs. Nat Commun 9(1):2186.

6. Bennett AF, Haslem A, Cheal DC, Clarke MF, Jones RN, Koehn JD, Lake PS, Lumsden LF, Lunt ID, Mackey BG, Nally RM (2009) Ecological processes: a key element in strategies for nature conservation. Ecol Manag & Restor 10(3):192–199
7. Bilkovic DM, Mitchell MM, Peyre L, Megan K, Toft JD, EBSCOhost (2017) Living shorelines: the science and management of nature-based coastal protection/edited by Donna Marie Bilkovic, Molly M. Mitchell, Megan K. La Peyre, Jason D. Toft. CRC Press, Boca Raton, FL
8. Cairns Regional Council (2016) CairnsPlan 2016 and Planning Scheme Policy—Natural Hazards. Available online from: https://www.cairns.qld.gov.au/property-and-business/planning-schemes/current/policies
9. Cairns Regional Council (2022) Cairns Climate Change Strategy 2030. Available online from: https://www.cairns.qld.gov.au/community-environment/climate-change/cairns-climate-change-strategy
10. Chen R, Zhang Y, Xu D, Liu, M (2017) Climate change and coastal megacities: Disaster risk assessment and responses in shanghai city. Climate Change, Extreme Events and Disaster Risk Reduction: Towards Sustainable Development Goals, 203–216
11. Cheshmehzangi A (2026) Thailand's tropical thresholds: coastal climate resilience and sustainable urban transformations. Springer, Singapore
12. Cheshmehzangi A, Alidoust S, Chen WY, Fuller R (eds) (2026) Designing Nature into Cities. Springer, Singapore
13. DCCEEW (2021) National climate resilience and adaptation strategy 2021–2025. Department of Climate Change, Energy, the Environment and Water, Australian Government. https://www.dcceew.gov.au/climate-change/adaptation/strategy
14. Das S, Dutta S, Roy J, Choudhury MR, Sen P, Chanda A, Dutta H, Nandi S, Ghosh T (2026) Mangroves in the Anthropocene: a global synthesis of carbon storage, biodiversity, and coastal resilience under climate and anthropogenic stressors. Sci Total Environ 1015:181405
15. Ferrario F, Beck MW, Storlazzi CD, Micheli F, Shepard CC, Airoldi L (2014) The effectiveness of coral reefs for coastal hazard risk reduction and adaptation. *Nat Commun* 5:3794
16. Fila D, Fünfgeld H, Dahlmann H (2024) Climate change adaptation with limited resources: adaptive capacity and action in small- and medium-sized municipalities. Environment, Development and Sustainability: A Multidisciplinary Approach to the Theory and Practice of Sustainable Development. Springer, 26(3), 5607–5627, March. https://doi.org/10.1007/s10668-023-02999-3
17. Fitton JM, Addo KA, Jayson-Quashigah PN, Nagy GJ, Gutiérrez O, Panario D, Carro I, Seijo L, Segura C, Verocai JE, Luoma S, Klein J, Zhang TT, Birchall J, Stempel P (2021) Challenges to climate change adaptation in coastal small towns: Examples from Ghana, Uruguay, Finland, Denmark, and Alaska. Ocean & Coastal Management, 212, Article 105787. https://doi.org/10.1016/j.ocecoaman.2021.105787
18. Goh K (2020) Flows in formation: the global-urban networks of climate change adaptation. Urban Stud 57(11):2222–2240
19. Government of the Cook Islands (2012) Cook Islands joint national action plan for disaster risk management and climate change adaptation (JNAP) 2011–2015. Office of the Prime Minister, Government of the Cook Islands
20. Government of the Cook Islands (2016) Marae Moana policy 2016–2020. Available online from: https://www.maraemoana.gov.ck/wp-content/uploads/2019/04/FINAL-Marae-Moana-Policy-2016-2020.pdf
21. Government of the Cook Islands (2018) Cook Islands climate change policy 2018–2028. Climate Change Cook Islands, Office of the Prime Minister. Available from: https://climatechange.gov.ck/wp-content/uploads/2019/10/Cook-Islands-Climate-Change-Policy-2018-2028.pdf
22. Government of Solomon Islands (2012) National climate change policy 2012–2017. Ministry of Environment, Climate Change, Disaster Management and Meteorology. Honiara, Solomon Islands
23. Grases A, Gracia V, García-León M, Lin-Ye J, Sierra JP (2020) Coastal flooding and erosion under a changing climate: implications at a low-lying coast (Ebro Delta). Water 12(2):346

24. Hoffmann TC (2002) The Reimplementation of the Ra'ui: coral reef management in Rarotonga, Cook Islands. Coast Manag 30(4):401–418. https://doi.org/10.1080/089207502900282
25. Hsiung AR, Hartanto RS, Bhatia N, Morris RL (2024) Challenges and opportunities for implementing nature-based coastal protection in an urbanised coastal city based on public perceptions. J Environ Manag 370:122620
26. Hülsen S, Dee LE, Kropf CM, Meiler S, Bresch DN (2025) Mangroves and their services are at risk from tropical cyclones and sea level rise under climate change. Commun Earth & Environ 6(1):262
27. IPCC (2021) Climate change 2021: the physical science basis. Cambridge University Press, Cambridge
28. Jansson L, Koschmieder A, Servières L et al (2025) Climate resilient urban development: an agenda for Small Island Developing States. npj Urban Sustain 5:99. https://doi.org/10.1038/s42949-025-00278-w
29. Jones TG, Glass L, Gandhi S, Ravaoarinorotsihoarana L, Carro A, Cripps G, Giri C (2016) Madagascar's mangroves: quantifying nation-wide and ecosystem-specific dynamics, and detailed contemporary mapping of distinct ecosystems. Remote Sens 8(2):106. https://doi.org/10.3390/rs8020106
30. Knutson T, Camargo SJ, Chan JC, Emanuel K, Ho CH, Kossin J, Mohapatra M, Satoh M, Sugi M, Walsh K, Wu L (2020) Tropical cyclones and climate change assessment: part II: projected response to anthropogenic warming. Bull Am Meteorol Soc 101(3):E303–E322
31. Li L, Cheshmehzangi A, Chan FKS, Ives CD (2021) Mapping the research landscape of nature-based solutions in urbanism. Sustainability 13(7):3876
32. Löf M, Madsen P, Metslaid M, Witzell J, Jacobs DF (2019) Restoring forests: regeneration and ecosystem function for the future. New For 50(2):139–151
33. Masson-Delmotte V, Zhai P, Pirani A, Connors SL, Péan C, Berger S, Caud N, Chen Y, Goldfarb L, Gomis MI, Huang M (2021) Climate change 2021: the physical science basis. Contribution of working group I to the sixth assessment report of the intergovernmental panel on climate change 2(1):2391
34. McHarg IL (1969) Design with nature. American Museum of Natural History, New York, pp 7–17
35. McIvor A, Tonneijk FH (2014) Mangroves for coastal defence. Guidelines for coastal managers & policy makers. *Published by Wetlands International and The Nature Conservancy*, 42
36. Mohd FA, Abdul Maulud KN, Karim OA, Begum RA, Awang NA, Hamid MA, Rahim NA, Razak AA (2018, June) Assessment of coastal inundation of low lying areas due to sea level rise. In *IOP conference series: earth and environmental science* (Vol. 169, No. 1, p 012046). IOP Publishing
37. Mori AS (2011) Ecosystem management based on natural disturbances: hierarchical context and non-equilibrium paradigm. J Appl Ecol 48(2):280–292
38. Morris R, Konlechner T, Ghisalberti M, Swearer S (2018) *From grey to green: efficacy of eco-engineering solutions for nature-based coastal defence.* Glob Chang Biol 24(5):1827–1842. https://doi.org/10.1111/gcb.14063
39. Morris RL, Pomeroy AWM, Boxshall A, Colleter G, Dack D, Dunlop AR, Hanslow D, King S, Magini A, O'Malley-Jones K, Sultmann S, Townsend M, Valesini F, White J, Zavadil E, Swearer SE (2024) A blueprint for overcoming barriers to the use of nature-based coastal protection in Australia. Front Environ Sci 12:Article 1435833. https://doi.org/10.3389/fenvs.2024.1435833
40. Narayan S, Beck MW, Reguero BG, Losada IJ, Van Wesenbeeck B, Pontee N et al (2016) The effectiveness, costs and coastal protection benefits of natural and nature-based defences. PloS one 11(5):e0154735
41. Nelson SH, Bigger P (2022) Infrastructural nature. Prog Hum Geogr 46(1):86–107. https://doi.org/10.1177/0309132521993916
42. Nicholls RJ, Wong PP, Burkett V, Codignotto J, Hay J, McLean R, Ragoonaden S, Woodroffe C (2007) Coastal systems and low-lying areas. Cambridge University Pres, Cambridge, pp 315–356. Available online from: https://pubs.usgs.gov/publication/70204340

43. Okamoto T, Doyon A (2024) *Equity and justice in urban coastal adaptation planning: New evaluation framework*. Build Cities 5(1):101–116. https://doi.org/10.5334/bc.377
44. Pelling M, Blackburn S (2014) Case studies: governing social and environmental transformation in coastal megacities. In: Pelling M, Blackburn S (eds) Megacities and the coast. Routledge, New York, pp 200–236
45. Remme D, Haarstad H (2022) *From instrumentalization to commoning: a critical review of participation in urban nature-based solutions*. Front Sustain Cities 4:917607. https://doi.org/10.3389/frsc.2022.917607
46. Rogers NJL, Adams VM, Byrne JA (2023) Factors affecting the mainstreaming of climate change adaptation in municipal policy and practice: a systematic review. Clim Policy 23(10):1327–1344. https://doi.org/10.1080/14693062.2023.2208098
47. Russo A, Baresi U, Cheshmehzangi A (2026) Nature-based solutions in urban regeneration: a review of methods, governance, and future directions. Urban Sci 10(3):130
48. Satterthwaite D, Archer D, Colenbrander S, Dodman D, Hardoy J, Mitlin D, Patel S (2018) Building resilience to climate change in informal settlements. One Earth 2(2):143–156
49. Seddon N, Chausson A, Berry P, Girardin C, Smith A, Turner B (2020) Understanding the value and limits of nature-based solutions to climate change and other global challenges. Philos Trans R Soc B 375:20190120
50. Silva R, Martínez ML, Odériz I, Mendoza E, Feagin RA (2016) Response of vegetated dune–beach systems to storm conditions. Coast Eng 109:53–62
51. Singh AR, Thirumurugan V, Bhomia RK, Prabakaran N (2024) Mangrove vegetation response to alteration in coastal geomorphology after an earthquake in Andaman Islands, India. RegNal Stud Mar Sci 76:103583
52. Temmerman S, Meire P, Bouma TJ, Herman P, Ysebaert T, De Vriend H (2013) Ecosystem-based coastal defence in the face of global change. Nature 504:79–83
53. UN-Habitat (2020) *World Cities Report 2020: The Value of Sustainable Urbanization*. Available online from: https://unhabitat.org/sites/default/files/2020/10/wcr_2020_report.pdf
54. United Nations Human Settlements Programme (2023) *Climate resilient Honiara*. Available online from: https://solomons.gov.sb/wp-content/uploads/2023/07/CRH-2023-summary.pdf
55. Wannewitz M, Ajibade I, Mach KJ et al (2024) Progress and gaps in climate change adaptation in coastal cities across the globe. Nat Cities 1:610–619. https://doi.org/10.1038/s44284-024-00106-9
56. Williams MJ, Coles R, Primavera JH (2007) A lesson from cyclone Larry: an untold story of the success of good coastal planning. Estuar Coast Shelf Sci 71(3–4):364–367. https://doi.org/10.1016/j.ecss.2006.09.010
57. Wolff E, Rauf HA, Hamel P (2023) *Nature-based solutions in informal settlements: A systematic review of projects in Southeast Asian and Pacific countries*. Environ Sci & Policy 145:275–285. https://doi.org/10.1016/j.envsci.2023.04.014
58. World Bank (2021) *A Catalogue of Nature-Based Solutions for Urban Resilience*. © World Bank. http://hdl.handle.net/10986/36507
59. Yang B, Li S (2016) Design with Nature: Ian McHarg's ecological wisdom as actionable and practical knowledge. Landsc Urban Plan 155:21–32
60. Zhang JE, Liu JL, Ouyang Y, Liao BW, Zhao BL (2010) Removal of nutrients and heavy metals from wastewater with mangrove Sonneratia apetala Buch-Ham. Ecol Eng 36(6):807–812

Interscaler Design for a Rewilded City: A Case Study of the Green Monterrey Metropolis

Rob Roggema and Igor Ishi Rubio Cisneros

1 Introduction

In 2014, the global population living in cities crossed the 50% boundary [97], and this will rise to 68% in 2050 [98]. Many urban residents live in vulnerable conditions, so it is necessary to have livable, clean, and healthy cities. Only 40% of urbanites have access to safe sanitation and drinking water, more than 90% are exposed to polluted air and many live close to the epicenters of (vector-borne) diseases [106]. Life in cities is impacted by climate change hazards, such as storms, droughts, and floods, which impact health, livelihoods, housing, and infrastructure [94]. At the same time, activities deployed in the city, e.g., transport and buildings, are responsible for 60–70% of carbon emissions [29, 33]. The role of urban development is therefore a potentially significant factor in reducing carbon emissions ([22]; Poumanyong & Kaneko [53, 58]). Especially in larger cities, the change from green and open surfaces to land uses of asphalt and concrete and the energy used in transport and buildings [34] are some of the root causes of the urban heat island. Moreover, in cities the majority of resource depletion is caused. The use of materials, energy, water, and food leads to (solid) waste [88], air pollution (Molina & Molina [45]), water pollution [12], and increased levels of emission [94]. Most of these impacts affect the human health of more vulnerable populations the most (EPA [24]). Urban development also influences the capacity of threatened ecosystems [102] that are subsequently degraded [114] and disrupt hydrological systems [43]. The urban footprint is a multiplier problem: ecosystems' health could provide healthier conditions for humans but their degradation makes this impossible. In this chapter, we present a spatial interscalar

R. Roggema (✉)
Cittaideale, Wageningen, The Netherlands
e-mail: rob@cittaideale.eu

I. I. R. Cisneros
Tecnologico de Monterrey, Monterrey, Mexico

A. Cheshmehzangi et al. (eds.), *Nature-based and Climate Responsive Design for Urban Systems*, Urban Sustainability, https://doi.org/10.1007/978-981-92-2070-0_8

strategy for how green spaces can reverse these conditions by connecting large-scale metropolitan design with the realization of hyperlocal projects.

2 Background Knowledge

2.1 Urban Problems

Some say that dense cities and metropolises are greener [55] and are more sustainable, e.g., reducing carbon emissions; these conurbations are not necessarily greener, ecological, or biophilic [14]. On the contrary, larger cities are less green (Oliveira et al. [52]). The current timeframe is a difficult one [61] because climate change is faster, less predictable, and possibly more impactful. Cities have not only become the cause of dangerous climate change, but they may also be carrying possible solutions for the reduction of harmful GHGs (UN News undated). One of the most serious problems emerging and accelerating due to urban activities is the degradation of ecosystems and biodiversity loss [38]. Other problems caused by and in cities are natural resources depletion [11, 56], human health impact [10], pollution [19], lack of spaces for physical health [25, 41], and mental health issues [39, 81]. Despite many cities aiming to green the city by deploying physical planning, experimenting with green infrastructure [9], only one-fifth embrace social and economic factors [27]. Green infrastructure is primarily used to link relict green areas to the built environment [1], thereby only partially resolving environmental problems. In some cases, a specific focus on urban ecology aims to '***support*** resilience-oriented development and highlight a synergetic response with an evolutionary form of multi-disciplinary sustainability' [83]. However, urban planners are constrained by limited knowledge and awareness in integrating this into their practices [101]. Research has repeatedly shown that urban ecology plays a beneficial role in improving cities' quality of life, yet it is not widely used in day-to-day planning processes for urban development.

In large cities, there is an urgent need to adapt to the predicted impacts of climate change. This is a spatial challenge [63] at the urban scale, but many policies focus on the national level. Approximately 15% of cities have presented initiatives and plans to adapt and only 18% plan for adaptation policies [8]. It is a challenge for cities that have developed within an industrial paradigm to envision a symbiotic future [57].

When people have access to green and nature, they generally feel happier [84]. According to the biophilia theory, natural environments are appreciated at a meta-level, higher than urban entities [108]. Nature positively influences mental fatigue, stress, and emotional well-being [28, 35, 93] through its fascination, novelty, coherence, and compatibility [35]. It can even have positive effects on hospital patients' recovery [92]. Traditional knowledge suggests that altered landscapes can heal and recover naturally when they are treated sustainably and made reciprocal and adaptable [107], in a way that benefits human health and, in turn, the landscape. This understanding of the reciprocity of human and their environment's health alludes to positive emotions dedicated to the landscape. In the Symbiocene human and non-human organisms will live in a sustainable equilibrium, not putting human pressure on the Earth system [3–7]. The words that are used to describe these emotions can be subdivided into three categories: pain, heal, and love ([80], Table 1):

- (Landscape) **Pain**—In the urban landscape, places, environmental qualities, spaces, and views can all hurt the person to whom it is exposed. These are visual, emotional, physical, and painful emotions, or related to a negative impact on people's health. The added visual, physical, or emotional painful experiences determine the landscape pain in an urban landscape.
- **Heal**—The identified pain can be resolved or healed. For a metropolis to heal itself, it requires a strong basic framework of eco-hydrological systems that create a large-scale healthy landscape, mitigating all forms of possible landscape pain. Healing is most effective at the larger spatial scale.
- **Love**—A basic healthy landscape is good for the overall quality of a region but when concrete projects are realized close to residential areas, so that people can experience and appreciate them. The realization of, for example, urban 'Miyawaki' forests or urban food gardens goes beyond repairing negative impacts or landscape pain. It adds to the empathy of people and to their care for their direct environment: people will start to fall in love with these places.

3 Materials and Methods

3.1 Case Study Area: Monterrey, Mexico

Monterrey is the second largest city in Mexico and is located some 200 km south of the American border, in the country's northeast. The southern and southwestern sides of the city are bounded by the Sierra Madre Oriental, a large mountain range that extends throughout the country. At the northeastern and eastern sides, the landscape transitions into the Tamaulipas arid plain heading to the Gulf of México. The Monterrey metropolitan region is the southernmost tip of the Rio Grande (or Rio Bravo) watershed, supplying its water to this system before it flows into the Gulf of Mexico. The city of Monterrey has expanded into the valley surrounded by mountains and hilly terrain (Fig. 1), which is why, in summer, heat can be trapped, in winter, combustion is trapped with the greenhouse effect, and, throughout the year, air pollution concentrates. The size of the Metropolitan of Monterrey extends roughly 70 x 70 km, and is inhabited by circa 5.3 million people (SEDATU et al. [86]). The foundations of Monterrey are laid by industries, such as Cemex (cement), Pemex (petrol), Femsa (soft drinks, beer, and food), and Ternium (steel), to name a few. They established the image of the city and the people's mentality too. The cities' environmental problems, such as water and air pollution,—quality and—availability, ecological degradation, and the unlimited urges of the residents to drive cars, have

Table 1 Identification of pain, heal, and love based on Earth emotions' vocabulary (after [5])

Verb	Translation	Etymology	Explanation
Pain			
Terraphthora	Earth-destroyer	Terra = earth; phthora = to destroy, destroy, ruin	Earth destroyers will be increasingly held accountable, both by the public and politicians
Tierratrauma	Earth trauma	Tierra = region/zone; trauma = wound	An 'acute earth-based existential trauma in the present'
Solastalgia	Consolation	Solari = to comfort; nostalgia = longing for past times	A 'pain or distress caused by the loss—Or lack of—Solace and the sense of desolation connected to the present state of one's home and territory'
Terrafuria	Climate rage	Terra = earth; furia = anger, rage	'The extreme anger unleashed within those who can see the self-destructive tendencies in the current forms of industrial-technological society and feel they must protest and act to change its direction'
Heal			
Sumbiofact	Built-with-nature	Sumbio = from the close cooperation between two or more different biological entities; factum = operation, creation, result of action/deed	Fabricated by human/nature interaction. As distinct from 'artifact', an object made artificially by a human being
Symbiocene		Symbiosis = living together; kainos = new	The Symbiocene will be in evidence when there is no discernible impact of human activity on the planet other than the temporary remains of their teeth and bones. Everything that humans do will be integrated within the support systems of all life and will leave no trace'
Love			
Endomophilia	Own-place-love	Endemos = native, at home, born and raised in, among us; philia = love for	'The particular love of that which is locally and regionally distinctive as felt by the people of that place'

(continued)

Table 1 (continued)

Verb	Translation	Etymology	Explanation
Eutierria	Planet connection	Eu = good; terra = earth; tierra = region/zone	'A positive and good feeling of oneness with the earth and its life forces where the boundaries between self and the rest of nature are obliterated and a deep sense of peace and connectedness pervades consciousness'
Psychoterratic	Earth soul movements	Psyche = the human soul or spirit; terratic = relating to the earth	'Emotions related to (positively and negatively) perceived and felt states of the earth'
Terranascia	Soil enricher	Terra = earth; nascia = to be born	'Green gold', 'earth creator'

been exaggerated by the rapid growth of the industries. The road system is overloaded with noise and air-polluting cars and trucks, which leaves a profound impression on city inhabitants and visitors alike.

Fig. 1 View over the metropolitan region of Monterrey from the Sierra Madre Oriental. *Source* picture taken by Rob Roggema

Table 2 Mapping of landscape pain and the sources used

Map	Data sources
Quarries and constructions in natural reserves	NPA, Open buildings dataset (Google Earth Engine), [89]
Large roads (more than six lanes each way)	[31]
XL and L buildings	Open buildings dataset (Google Earth Engine), [89]
Heat islands	[21]
Railways and above metro system	[32], PRIMUS 2019
Energy infrastructure/air quality	[26, 51]
Hydrological obstructions	[32, 54]
Landscape transformation	[13, 17]

3.2 Methods

The research has been carried out using three different methods: Mapping (landscape pain), Landscape First (to heal), and design, plan, and realize (hyperlocal projects).

3.2.1 Mapping Landscape Pain

To identify landscape pain in the region, a series of mappings has been undertaken. The most pronounced impacts on the experience of the city were selected first. After checking this selection with the client (Fondacion Ambiental de Monterrey Metropolitana), the final list of landscape pain topics was adopted. The most significant ones are infrastructural barriers, extra-large buildings, impaired mountains and natural reserves, air pollution, and disrupted waterways. These topics were then mapped in separate maps and integrated into one landscape pain map, and one hydrological pain map (due to the vast amount of data found). The selected maps and the sources that were used to create them are summarized in Table 2.

The maps of the Monterrey Metropolitan Region show the abundance of the natural landscape surrounding the city, including its rivers and creeks, forests, and mountains. However, these natural qualities are hugely impacted by human activities and industries. Where the maps of the natural conditions show a great interdependency between soil, elevation, ecology, and hydrology, the urban maps provide a different image of urban heat islands, air pollution, and a deficit of green spaces.

3.2.2 'Landscape First' to Heal the Landscape

In a research-by-design process [67] different strategies to heal the landscape pain have been revealed. Beyond more traditional scenario planning [60, 74, 100, 110], the real connection to the eco-hydrological network is made by applying a Landscape First method. In this landscape-based regional design [47], the physical form of regions is shaped by using the landscape as the basic condition [48]. The biosphere is seen as the context for long-term urban development [49, 50], identifying a common orientation for the spatial decisions and aiming to create inspiring landscape ecologies [75] by embedding natural resources within the urban landscape [72]. This way, hidden landscape-ecologies can be brought to the surface again [76] regenerating the land by using historic landscape dynamics [69, 77].

3.2.3 Realized Projects to Love the Environment

To connect the large-scale analysis of landscape pain and the systems approach to healing, tactics of hyperlocal projects are deployed in parallel. These tactics contain concrete realized projects that illuminate the benefits of green, accessible, and beautiful spaces [73]. At this scale and during the regional design research, two concrete projects have been realized: the Urban Milpa and two Miyawaki forests.

The Milpa system is a traditional Mesoamerican and Mexican method of cultivating food in a symbiotic manner. It has long-standing traditions and is a good example of a sustainable and biodiverse approach to local food production [15]. It is normally practiced in rural areas and small towns. To bring traditional recipes, cultivation, and consumption closer to residents, northeastern recipes have been investigated to identify specific crops for cultivation [78, 79]. These crops are subsequently grown in an Urban Milpa. This specific form of milpa enables cultivation in urban conditions. Raised plantbeds are constructed and positioned in a way that they can be placed in any small urban space, such as a garden, a front yard, or a small wasteland. Each year, two cycles of planting to harvesting are possible in the urban milpa, and the harvest is given to local chefs who adjust the traditional recipes to current times and cook the dishes for a selected group of local people.

The second hyperlocal project that is realized in parallel with the regional analysis and design is the Miyawaki forest. This specific method of planting small forests in high densities was developed by prof Miyawaki [44]. It is a way of bringing local plantations very close to the urban residents [70]. The Miyawaki method aims to plant trees very closely together, so they start competing with each other and growing faster than other regular forests [2, 59]. It is expected that these forests yield better environmental results, such as more than 30% higher capture of GHGs, over 300 new species, less noise and air pollution, cooler temperatures, and more humid conditions [46, 91]. In Monterrey, two of these forests have been planted (in October 2024 and February 2025). The forests are monitored using sensors, and the number and type of bird species are also tracked using the Haikubox [42].

4 Results

4.1 Landscape Pain Is Widespread

Figure 2 shows the integration of a series of maps with analyses of the topics selected as potentially landscape pain. The following maps have been constructed:

- Quarries and buildings in nature reserves: 1289 ha of quarries, 29,928 buildings on mountain slopes (3497 in Natural Protected Areas).
- Green spaces: 22.2 m^2 per inhabitant. This is nearly 30 m^2 less than in Barcelona [30].
- Urban green space: 4 m^2 per capita. WHO advises a minimum of 9.5 m^2 [104]. 30% green cover is realized only in 50% of the region, and only 43% has a green park within a 300-m walk [105]. With these numbers, the 3–30-300 rule [37] is not met in the region of Monterrey. This lack of green cover and spaces increases the exposure of especially vulnerable people to urban heat [113].
- 670 km of main roads, over six lanes and 996 ha of parking. Carbon emissions by cars using this infrastructure: 7,764,568 tons of CO_2/year. It impacts air quality [40], soil quality [87], and urban heat [109].
- In the Monterrey region, there are 972 buildings larger than 10,000 m^2 (XL) and 7628 between 2500 and 10,000 m^2. With their dominant paved surfaces and roofs, they contribute considerably to the urban heat island effect [62, 85]. Moreover, visually they are generally not appreciated.
- 411 km of rail tracks create a severe noise problem of 80–100 dB [18] in the city. Together with the elevated metrolines (19.3 km and expanding), they also form spatial barriers between neighborhoods, dangerous underpasses, and block views of the surrounding landscape.
- Emission of PM2.5 and PM10, mainly by the larger industries, impacts human health and life-threatening diseases, leading to 1684 deaths in 2021. Polluting industries such as the Pemex refinery emit the majority of PM10 and PM2.5 particles, causing adverse effects on human health and life-threatening diseases [36].

When all landscape pain is collected and integrated in one map (Fig. 2), a strong correlation becomes visible between concentrations of oversized buildings, low percentages of green cover, urban heat, and air pollution, leading to adverse effects on ecological and human health.

Similar to the analysis of landscape pain, hydrological pain is also apparent in the Monterrey region (Fig. 3). The natural flow of rivers (rio's) and creeks (arroyo's) is seriously disturbed in 17,003 places. Flow is interrupted by overbuilding creeks with houses or roads; formerly, natural creeks have become part of the sewage, and water is invisible, beneath pavement. Where the water disappeared, the highest risk of flooding appears, and the lowest water quality and least ecological quality are found [23].

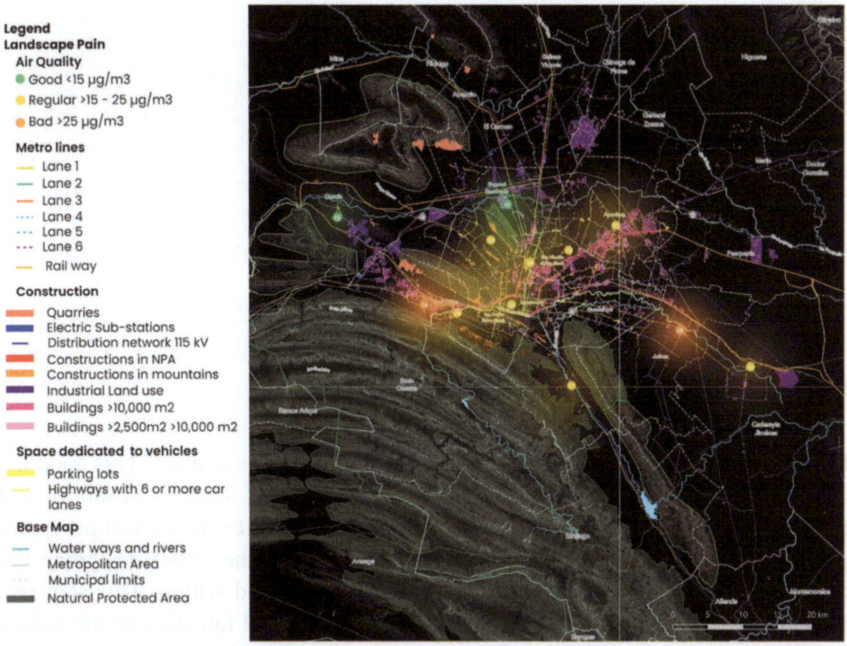

Fig. 2 Collated landscape pain in Monterrey. (*Source* [80])

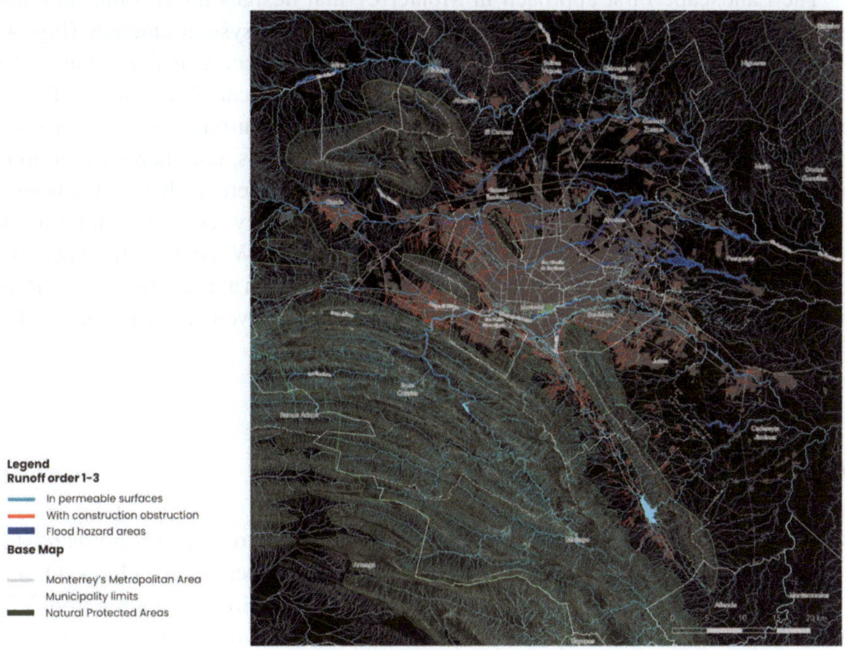

Fig. 3 Hydrological pain in Monterrey. *Source* [80]

Fig. 4 Landscape first approach applied to the Monterrey region. *Source* [71]

4.2 Healing

To heal the landscape pain, the Landscape First approach is applied. This approach starts to analyze the natural conditions of the landscape, such as elevation, soil, hydrology, and ecology, and based on these elements, it creates a natural spatial green framework. This is established first, hence the name. After designing this spatial framework, urban and other uses can be embedded within. This sequence of planning decisions guarantees that the functioning and qualities of the natural systems prevail over other ones, hence creating a healthy urban condition for urban life.

The Landscape First approach in Monterrey first designs the reconnection and uncovering of all waterways so a coherent hydrological system emerges (Fig. 4). Added to this blue network is a robust system of green eco-corridors, that widen the waterways significantly. In these green zones, water can flow when it floods, ecosystems flourish, and people can recreate. Once this green-blue system is set, other urban uses can be embedded, such as housing, workplaces, commercial functions, and infrastructure. These should not destabilize or interfere with the functioning of the eco-hydrological system. The plan for the Monterrey region is similar to the landscape-based structure plan designed in the 1960s [20]. When the natural network is reestablished, it brings many environmental benefits, such as reduction of the urban heat island, increased humidity in dry contexts and improvement of the air quality [82].

4.3 Love

As many emotions, love is also subjective. You can feel it, or not. When it is related to spatial interventions, the new urban form can create affectionate feelings. Or not. In general, urban residents feel more connected to their direct living environment. Therefore, when the landscape pain is healed, the focus lies on the overall conditions for a healthy urban landscape, while the appreciation, or love for the environment is found in realizing hyperlocal projects.

Fig. 5 The first Miyawaki forest at its plantation (30 Oct 2024) and 10 months later (20 August 2025). *Source* pictures taken by Rob Roggema

4.3.1 Miyawaki Forest

Two Miyawaki forests are realized, one in October 2024 and one in February 2025. Both are realized within the premises of the Tecnologico de Monterrey campus. For the planting of the first forest, approximately 45 people participated. These included students, faculty of the university, artists, friends, and representatives of NGOs (i.e. http://www.marsustentable.org/) and industry from the region. Since its plantation, the forest has grown very well (Fig. 5). The planting of the forest created a tense atmosphere among the planters, who immediately began talking with one another, making plans to return and contemplate near the forest, and could not stop taking pictures of themselves planting trees. In the months after, many initial planters returned to the forest, just to stop by, sit and think, or make pictures of the growth of the forest. This reflects a positive sentiment toward the place being transformed into a small forest. It also means that this place represents for many the desired transition from a painful urban landscape toward a regenerated green metropolis.

The second forest was planted in a similar atmosphere of togetherness and emotional connectedness. This time, approximately 35 planters were assisting, of whom most had not participated in the first plantation. A group of Faculty of Excellence professors, faculty, family members, and students participated this time. The responses were the same: positivism, a representation of a greater goal, and the desire to return to this place many times in the future. Several professors proposed establishing similar forests on their campuses.

4.3.2 Urban Milpa

For the construction and three growth and harvest cycles, a small team of 3–4 people managed the construction of the raised beds, the soil preparation, the seeding maintenance, and the facilitation of the growth of the crops (Fig. 6). On a weekly basis, the plants are watered, and checks are carried out to control eventual pests. After several months, the Squash, Corn, and Beans can be harvested. Each produce is measured and weighed. After the harvest is completed, the crops are delivered to the chef, in the first cycle, to Rodrigo Rivera, chef of KOLI (https://koli.mx). He then used the produce to reinvent traditional dishes and cooked them for the project team (Fig. 7). The taste is amazing, and the people enjoyed the process of constructing the urban milpa, the care for the plants, the harvest, and finally, a tasting event. The people involved were emotionally connected to the growing process and the transformation of a small piece of land into a productive urban site. The urban milpa connected them to the bigger regional and local food system. The team continued to work on the second growth cycle, with similar results. Instead of kalebas and garlic, tomatoes, and acanthus were grown. A second tasting event took place in Vernáculo (https://vernaculo.com/), and chef Hugo Guajardo created the dishes with the crops provided.

The urban milpa example shows that a small food garden can make a significant difference, crops can be grown in small plots, and traditional recipes can be transformed into modern, tasty dishes. This gives confidence that the small experiment can be scaled up to other locations within the Monterrey metropolis, especially where these gardens can contribute to improving the health of vulnerable communities. The expansion of urban milpas throughout the city contributes to the metropolitan green and blue network. It represents the first little steps in transforming the entire metropolis into a green one.

Fig. 6 The urban milpa from construction to (almost) harvest time: 15 August–15 October 2024

Fig. 7 Tasting event at restaurant La Cometa (Rodrigo Rivera)

5 Discussion and Implications

5.1 Discussion

There are many cities that have profound greening plans, in response to global pledges for naturalizing the urban environment [96, 103]. Soma has a focus on water [112], nature [16], or trees [90], but they all aim to counteract economically driven urban development in some way. To a certain extent, these plans and programs improve the overall environmental quality and biodiversity in cities. However, they have a limited impact when compared to the ongoing urban growth in housing, industries, traffic, and infrastructure. Therefore, all efforts that are put into greening the city and hence increasing its resilience are counteracted by an urban economics push.

The main flaw is that the city's development is driven primarily by its urban uses. It starts with the need for housing, workplaces, malls and shops, and mobility. At a later stage in the planning and design process, the other aspects that make the city livable are introduced: leisure and recreation, water management, ecology, and beauty. Although the policy and planning documents are full of green ambitions and paragraphs of a sustainable future, in reality, these objectives are often degraded to 'nice to haves' when they compete with the perceived 'need to haves'.

The only way to lay a robust foundation for a nature-driven city is to start with the landscape in every urban development process. Not only in the form of an analytical tool, but mainly as the spatial foundation for the entire metropolitan area. The eco-hydrological systems will then form a connected systemic network that can function optimally under changing conditions around and in between them. Urban use can be embedded in this network without compromising it.

Although limited in number, previous studies have explored the landscape-first approach. However, many of these plans and projects are developed at the local or subregional landscape scale. The Floodable Eemsdelta project [64], for example, analyzes the current elevation, hydrology, and ecology, before designing an anticipative plan for a potential dike breach and flood. The Mother Zernike project [69, 77] starts a long-term spatial vision in the mechanisms of the landscape, dealing with soil subsidence, sea level rise, the controlled water system, and a degraded ecology, transforming the agricultural system in a nature-rich landscape that is growing along with the pace of sea level rise, and a saline type of economically future-proof agriculture.

Other projects include the Westerpark neighborhood in Breda [68], in which the local catchment has been taken as the driver for the urban design, the tsunami recovery designs for Kessenuma [111] and Minamisoma [66] in Japan, and the bushfire resilient landscapes of Murrindindi and Bendigo in Victoria, Australia [65].

5.2 Implications

The consequences of such a different, landscape-oriented, approaches are diverse.

First of all, it requires a change of mindset. (Urban) Designers, decision-makers, residents, and everyone who is potentially involved in city-making should shift from viewing the city as a machine that is providing human beings with what they want, toward the city as a living system, in which the human demands are simply a part of the wider natural world.

Secondly, urban planning processes should prioritize the eco-hydrological system at every step of the design process, beginning at the level of the metropolitan (bio)region.

Finally, the implication of creating a well-functioning eco-hydrological urban network is that it will need to be supported by (hyper)local projects that are realized in line with the bigger system, at the same time providing residents with the healthy environment they deserve.

6 Concluding Remarks

In this chapter, we have seen that current greening strategies are not sufficient to create cities that are healthy for their human and non-human inhabitants. We propose a strategic methodology to identify the main problems (pain), ways to solve these (heal) and transform areas into healthy places (love). The urban landscape pain primarily focuses on problems that are experienced by the residents in the city. The mapped inventory of this landscape pain gives us an integrated view of all interlinked urban problems. One strategy that can help solve the landscape pain is healing. It is oriented toward a systemic approach to improving the health of eco-hydrological systems, thereby mitigating local problems. Within this healed landscape, specific places and areas are subsequently transformed into places that are embraced by its residents. Places they will care for and love.

The principle of Landscape First supports these traits of methodological steps. This means that in every design and planning process the landscape and its eco-hydrological features are analyzed and designed first. This is an essential choice to improve the overall health of the urban system. Once the natural network is established and can function in its natural way, the resilience of the entire area is large. Future shocks, disruptions and uncertainties can be mitigated without compromising the functioning of the natural system itself. This is to the benefit of the urban system.

This study is of great importance to other large urban regions worldwide. Many metropolises suffer from the urban heat island effect, flash flooding resulting from a high percentage of impermeable surfaces, ecological degradation, reduced area, and air (water and soil) pollution. The presented Green Metropolis can serve as a methodological approach applicable to all types of large urban regions. The search for the unique and local landscape pain that requires a specific response from an eco-hydrological perspective to create a regenerative system, and the specific local interventions that realize (parts of) the eco-hydrological framework, can be applied to these other areas. It is recommended to test and experiment with this methodology to tailor the approach and adapt to the local conditions.

Finally, the design strategy for larger urban regions/metropolises connects the large-scale and long-term systemic perspective with the hyperlocal realization of projects. As we have seen, the recovery and reconnection of the hydrological system in the metropolis of Monterrey are supported by individual, smaller-scale projects, such as the planting of Miyawaki forests or the cultivation of local food gardens. This is not only from a green-ecological perspective, but also from a social standpoint. When urban residents experience the transformation firsthand, they are more likely to take care of their environment. This may take time, but eventually the transformation of larger urban regions into a green wave becomes a reality.

References

1. Ahern J (2007) Green infrastructure for cities: the spatial dimension. In: Novotny V, Brown P (eds) Cities of the future towards integrated sustainable water and landscape management. IWA Publishing, London
2. Akram M, Khan M, Nabi T, Qadri R, Al-Maskri A, Khan MA (2025) Miyawaki technique for sustainable urban greening and ecological restoration: A review 21:0028. https://doi.org/10.1079/cabireviews.2025.0028
3. Albrecht G (2014) Ecopsychology in the symbio-cene Ecopsychology 6:58–59
4. Albrecht G (2015) Exiting the Anthropocene and Entering the Symbiocene. https://glennaalbrecht.wordpress.com/2015/12/17/exiting-the-anthropocene-and-entering-the-symbiocene/
5. Albrecht GA (2019) Earth emotions: new words for a New World. Cornell University Press, Ithaka, NY
6. Albrecht GA (2020) Negating solastalgia: an emotional revolution from the Anthropocene to the Symbiocene. Am Imago 77:9–30
7. Albrecht G, Sartore GM, Connor L, Higginbotham N, Freeman S, Kelly B, Stain H, Tonna A, Pollard G (2007) Solastalgia: the distress caused by environmental change. Australas Psychiatry 15(Suppl. 1):S95–S98
8. Araos M, Berrang-Ford L, Ford JD, Austin SE, Biesbroek R, Lesnikowski A (2016) Climate change adaptation planning in large cities: a systematic global assessment. Environ Sci Pol 66:375–382
9. Ashinze UK, Edeigba BA, Umoh AA, Biu PW, Daraojimba AI (2024) Urban green infrastructure and its role in sustainable cities: a comprehensive review. World J Adv Res Rev 21:928–936
10. Ashley J, Zhou B, Barry M (2017) The impact of megacities on health: preparing for a resilient future. Lancet Planet Health 1:e176–e178
11. Bahers J-B, Rosado L (2023) The material footprints of cities and importance of resource use indicators for urban circular economy policies: a comparison of urban metabolisms of Nantes-Saint-Nazaire and Gothenburg. Clean Prod Lett 4:100029
12. Bakács T (2013) Public health problems in metropolitan areas. In: Miles S (ed) Metropolitan Problems. Routledge, London, pp 191–213
13. Balon N, Hrdalo I, Mrda A, Kamenečki M, Tomić Reljić D, Pereković P (2023) Landscape urbanism—retrospective on development, basic principles and application. Architecture 3:739–752
14. Beatley T (2010) Green Metropolis. A new book argues that Manhattan is the greenest city in the United States. Places J
15. Benrey B, Bustos-Segura C, Grof-Tisza P (2024) The mesoamerican milpa system: traditional practices, sustainability, biodiversity, and pest control. Biol Control 198:105637. https://doi.org/10.1016/j.biocontrol.2024.105637

16. C40 & ARUP (2023) Urban Rewilding. The Value and co-Benefits of Nature in Urban Spaces. https://c40.my.salesforce.com/sfc/p/#36000001Enhz/a/Hp000000haJF/GiSUGhj3_IYcI_fm4n08xW8Vap411v2vVsjom_rGzCw
17. Cao Q, Liu Y, Georgescu M, Wu J (2020) Impacts of landscape changes on local and regional climate: a systematic review. Landsc Ecol 35:1269–1290
18. CDC (Center for Disease Control and Prevention) (2020) What Noises Cause Hearing Loss? www.cdc.gov/nceh/hearing_loss/what_noises_cause_hearing_loss.html
19. Cerón Breton RM, Céron Breton J, la Luz D, Espinosa Fuentes M, Kahl J, Espinosa Guzman AA, Martínez RG, Guarnaccia C, Del Carmen Lara Severino R, Ramirez Lara E, Francavilla AB (2021) Short-term associations between morbidity and air pollution in metropolitan area of Monterrey. Mexico Atmosphere 12:1352
20. Cortés-Melo G (2014) La Planificación de una Metrópoli: Por un Urbanismo Integral, Humanista y Sustentable, Tomo I. UANL, Monterrey. p 272
21. Debbage N, Marshall Shepherd J (2015) The urban heat Island effect and city contiguity. Comput Environ Urban Syst 54:181–194
22. Dhakal S (2008) Climate change and cities: the making of a climate friendly future. In: Droege P (ed) Urban energy transition: from fossil fuel to renewable power. Elsevier, Oxford, UK, pp 173–182
23. Ellis JB (1989) The management and control of urban runoff quality. Water Environ J 3:116–124
24. EPA (2021) Climate change and social vulnerability in the United States: A Focus on Six Impacts (EPA 430-R-21-003). https://www.epa.gov/cira/social-vulnerability-report
25. Frank LD, Engelke PO (2001) The built environment and human activity patterns: exploring the impacts of urban form on public health. J Plan Lit 16:202–218
26. Gobierno del Estado de Nuevo León (2018) Memorias del 1er Encuentro Energético Nuevo León 2018. https://www.nl.gob.mx/es/publicaciones/memorias-del-1er-encuentro-energetico-nuevo-leon-2018
27. Hanna E, Comín FA (2021) Urban green infrastructure and sustainable development: a review. Sustainability 13:11498
28. Hartig T, Evans GW, Jamner LD, Davis DS, Gärling T (2003) Tracking restoration in natural and urban field settings. J Environ Psychol 23:109–123
29. IEA (2008) World energy outlook. https://www.iea.org/reports/world-energy-outlook-2008
30. INE & IEC (2022) Instituto Nacional d'Estadística i Institut d'Estadística de Catalunya. https://www.idescat.cat/emex/?id=080193&lang=es
31. INEGI (2020a) Simulador de Flujos de Agua de Cuencas Hidrográficas (SIATL). https://antares.inegi.org.mx/analisis/red_hidro/siatl/#
32. INEGI (2020b) Censo Poblacional. www.inegi.org.mx/programas/ccpv/2020/#resultados_generales
33. IPCC (2022) Summary for policymakers. In: Pörtner H-O, Roberts DC, Tignor M, Poloczanska ES, Mintenbeck K, Alegría A, Craig M, Langsdorf S, Löschke S, Möller V et al (eds) Climate change 2022: impacts, adaptation, and vulnerability; contribution of working group II to the sixth assessment report of the intergovernmental panel on climate change. Cambridge University Press, Cambridge UK, New York, NY, pp 3–33
34. Jabbar HK, Hamoodi M, Al-Hameedawi AN (2023) Urban heat islands: a review of contributing factors, effects and data. IOP Conf Ser Earth Environ Sci 1129:012038
35. Kaplan R, Kaplan S (1989) The experience of nature: a psychological perspective. Cambridge University Press, New York, NY
36. Kheifets L, Swanson J (2020) Electric and magnetic fields and cancer: the epidemiological evidence. Environ Res 186:109565
37. Konijnendijk CC (2023) Evidence-based guidelines for greener, healthier, more resilient neighborhoods: introducing the 3–30–300 rule. J For Res 34:821–830
38. Kowarik I (2011) Novel urban ecosystems, biodiversity, and conservation. Environ Pollut 159:1974–1983

39. Krefis AC, Augustin M, Schlünzen KH, Oßenbrügge J, Augustin J (2018) How does the urban environment affect health and Well-being? A systematic review Urban Sci 2:21
40. Litman T (2013) Transportation and public health. Annu Rev Public Health 34:217–233
41. Lopez RP, Hynes HP (2006) Obesity, physical activity, and the urban environment: public health research needs. Environ Health 5:1–10
42. Mann D, Anderson A, Donner A, Hall M, Kahl S, Klinck H (2025) Continental-scale behavioral response of birds to a total solar eclipse. Sci Rep 15(1):1–10
43. McGrane SJ (2016) Impacts of urbanisation on hydrological and water quality dynamics, and urban water management: a review. Hydrol Sci J 61:2295–2311
44. Miyawaki A, (1993) Restoration of native forests from Japan to Malaysia. In: Lieth H, Lohmann M (eds.) Restoration of tropical forest ecosystems. Tasks for vegetation science. 30. Springer, Dordrecht. https://doi.org/10.1007/978-94-017-2896-6_1
45. Molina MJ, Molina LT (2004) Megacities and atmospheric pollution. J Air Waste Manag Assoc 54:644–680
46. Nargi L (2019) The Miyawaki method: a better way to build forests? https://daily.jstor.org/the-miyawaki-method-a-better-way-to-build-forests/
47. Nikologianni A, Moore K, Larkham PJ (2019) Making sustainable regional design strategies successful. Sustainability 11:1024
48. Nijhuis S (2019) Cultivating regions through design. Atlantis 29:43–46
49. Nijhuis S (2022) Landscape-based urbanism: cultivating urban landscapes through design. In: Roggema R (ed) Design for regenerative cities and landscapes: rebalancing human impact and natural environment. Springer, Dordrecht, pp 249–277
50. Nijhuis S, Xiong L, Cannatella D (2022) Towards a landscape-based regional design approach for adaptive transformation in urbanizing deltas. Res Urban Ser 6:55–80
51. Observatorio de la Calidad del Aire (2024) Datos históricos sobre la calidad del aire en el Área Metropolitana de Monterrey. Instituto Nacional de Ecología y Cambio Climático, México City
52. Oliveira EA, Andrade JS Jr, Makse HA (2014) Large cities are less green. Sci Rep 4:4235
53. O'Neill BC, Dalton M, Fuchs R, Jiang L, Pachauri S, Zigova K (2010) Global demographic trends and future carbon emissions. Proc Natl Acad Sci 107:17521–17526
54. Open Buildings Dataset (n.d.) (Google Earth Engine) https://sites.research.google/gr/open-buildings/#open-buildings-download
55. Owen D (2009) Green Metropolis: why living smaller, living closer, and driving less are the keys to sustainability. Riverhead Books, New York NY
56. Papa R, Gargiulo C, Zucaro F, Angiello G, Carpentieri G (2014) Urban energy consumptions: its determinants and future research. WIT Trans Ecol Environ 191:561–570
57. Pasini R (2017) The symbiotic field. Landscape paradigms and post-urban spaces; The Urban Book Series. Springer, Cham
58. Poumanyvong P, Kaneko S (2010) Does urbanization lead to less energy use and lower CO_2 emissions? A cross-country analysis. EcolEcon 70:434–444
59. Qi H, Dempsey N, Cameron R (2024) Seeing the forest for the trees? An exploration of the Miyawaki forest method in the UK. Arboricultural Journal 46(4):292–304. https://doi.org/10.1080/03071375.2024.2394355
60. Ramirez R, Selsky JW, Van der Heijden K (2010) Business planning for turbulent times: new methods for applying scenarios. Routledge, Abingdon
61. Ripple WJ, Wolf C, Gregg JW, Rockström J, Mann ME, Oreskes N, Lenton TM, Rahmstorf S, Newsome TM, Xu C et al (2024) The 2024 state of the climate report: perilous times on planet earth. Bioscience 74:biae087
62. Rizwan AM, Dennis LYC, Liu CA (2008) Review on the generation, determination and mitigation of urban heat Island. J Environ Sci 20:120–128
63. Roggema R (2009) Adaptation to climate change: a spatial challenge. Springer, Dordrecht
64. Roggema R (2012a) Swarm planning: the development of a methodology to Deal with climate adaptation. Delft, Wageningen: Delft University of Technology and Wageningen University and Research Centre. PhD-thesis

65. Roggema R (2012b) Swarming landscapes. In: Roggema R (ed) *Swarming landscapes: the art of designing for climate adaptation.* Springer, Dordrecht, Heidelberg, London, pp 167–193
66. Roggema R, Middleton L, Yan W (2017) Planning and design in Minamisoma: reborn, rethink, return. In: Roggema R, Yan W (eds) Tsunami and Fukushima disaster: design for reconstruction. Springer, Dordrecht, Heidelberg, London, pp 63–94
67. Roggema R (2017) Research by design: proposition for a methodological approach. Urban Sci 1(1):2. https://doi.org/10.3390/urbansci1010002
68. Roggema R (2020) Urbanism on water and ecology: the early example of Westerpark, Breda. In: Roggema R (Ed.) Contemporary urban design thinking, Vol. 2, Nature Driven Urbanism. Book series. Springer, pp 155–174
69. Roggema R (2021) From nature-based to nature-driven: landscape first for the design of Moeder Zernike. Sustainability 13:2368
70. Roggema R (2024a) Let the trees ´talk´: giving voice to nature through an immersive experience. WORLD 5:313–324. https://doi.org/10.3390/world5020017
71. Roggema R (2024b) From your balcony to the world: an Ecocathedric approach to regenerating urban landscapes; Keynote at the Foro de Innovación y Sostenibilidad Urbana, San Pedro Garza Garcia
72. Roggema R, Monti S (2021) Nature driven planning for the FEW-nexus in Western Sydney. In: Roggema R (ed) TransFEWmation: towards design-led food-energy-water Systems for future urbanization. Springer, Dordrecht, pp 59–94
73. Roggema R, Tillie N (2022) Realizing emergent ecologies: nature-based solutions from design to implementation. Land 11:1972. https://doi.org/10.3390/land11111972
74. Roggema R, Dassen T (2023) Thinking in improbabilities. In: Roggema R (ed) Trends in urban design: insights for the future urban professional. Springer, Dordrecht, pp 69–79
75. Roggema R, Keeffe G, Tillie N (2021a) Nature-based urbanization: scan opportunities, determine directions and create inspiring ecologies. Land 10:651
76. Roggema R, Tillie N, Keeffe G, Yan W (2021b) Nature-based deployment strategies for multiple paces of change: the case of Oimachi. Japan Urban Plan 6:143–161
77. Roggema R, Tillie N, Hollanders M (2021c) Designing the adaptive landscape: leapfrogging stacked vulnerabilities. Land 10:158
78. Roggema R, Mallet AE, Krstikj A (2023) Creating a virtuous food cycle in Monterrey. Mexico Sustainability 15(10):7858. https://doi.org/10.3390/su15107858
79. Roggema R, Krstikj A, Flores B (2024) Spatial barriers to transforming toward a healthy food system in the Noreste of Mexico. Nutrients 16:1259. https://doi.org/10.3390/nu16091259
80. Roggema R, Rubio Cisneros II, Junco López R, Ramirez Leal P, Ramirez Suarez M, Ortiz Díaz M (2025) Loving and healing a hurt city: planning a green Monterrey metropolitan area. Land 14:164. https://doi.org/10.3390/land14010164
81. Rose N (2020) Mental health, stress and the contemporary metropolis. In: Keith M, De Souza Santos AA (eds) Urban transformations and public health in the Emergent City. Manchester University Press, pp 35–54
82. Rubio Cisneros II (2023) Preliminary cartographic diagnosis for environmental and geological compensations in the municipality of Guadalupe, Nuevo León. Ing Investig Tecnol XXIV:1–8
83. Russo A, Cirella GT (2020) Urban sustainability: integrating ecology in city design and planning. In: Cirella GT (ed) Sustainable human–nature relations: environmental scholarship, economic evaluation, urban strategies. Springer, Singapore, pp 187–204
84. Samavati S, Veenhoven R (2024) Happiness in urban environments: what we know and don't know yet. J Housing Built Environ 39:1649–1707
85. Santamouris M (2015) Analyzing the heat Island magnitude and characteristics in one hundred Asian and Australian cities and regions. Sci Total Environ 512–513:582–598
86. SEDATU CONAPO INEGI (2024) Metrópolis de Mexico 2020. Gobierno de México: México City. https://www.gob.mx/cms/uploads/sedatu/MM2020_06022024.pdf
87. Shoub D (2011) The high cost of free parking. Planner Press, Chicago, IL
88. Siddiqua A, Hahladakis JN, Al-Attiya WAK (2022) An overview of the environmental pollution and health effects associated with waste landfilling and open dumping. Environ Sci Pollut Res 29:58514–58536

89. Sirko W, Kashubin S, Ritter M, Annkah A, Bouchareb YSE, Dauphin Y, Keysers D, Neumann M, Cisse M, Quinn JA (2021) Continental-scale building detection from high resolution satellite imagery arXiv arXiv: 2107.12283
90. Trees of the World (n.d. undated) https://treecitiesoftheworld.org
91. Treviño R (2024) The Miyawaki method: an urban solution to combat deforestation. https://tecscience.tec.mx/en/human-social/tiny-forest/
92. Ulrich RS (1984) View through a window may influence recovery from surgery. Science 224:224–225
93. Ulrich RS, Simons RF, Losito BD, Fiorito E (1991) Stress recovery during exposure to natural and urban environments. J Environ Psychol 11:201–230
94. UNEP (n.d.-a undated) Cities and climate change. https://www.unep.org/explore-topics/resource-efficiency/what-we-do/cities-and-climate-change#:~:text=Rising%20global%20temperatures%20causes%20sea,housing,%20human%20livelihoods%20and%20health
95. UNEP (n.d.-b undated) Sustainable, resource efficient cities—making it happen! https://sustainabledevelopment.un.org/content/documents/1124SustainableResourceEfficientCities.pdf
96. UN Habitat (2016) Greener cities partnership. https://unhabitat.org/greener-cities-partnership
97. United Nations (2015) World urbanization prospects: The 2014 Revision, (ST/ESA/SER.A/366). https://www.un.org/development/desa/pd/sites/www.un.org.development.desa.pd/files/files/documents/2020/Oct/unpd_2014_wup-report.pdf
98. United Nations (2019) World urbanization prospects: The 2018 Revision (ST/ESA/SER.A/420). https://population.un.org/wup/assets/WUP2018-Report.pdf
99. UN News Cities (n.d. undated) A 'Cause of and Solution to' Climate Change. https://news.un.org/en/story/2019/09/1046662
100. Van der Heijden K (2005) Scenarios: the art of strategic conversation. Wiley, Hoboken, NJ
101. Van Zyl B, Cilliers EJ, Lategan LG, Cilliers SS (2021) Closing the gap between urban planning and urban ecology: a south African perspective. Urban Plan 6:122–134
102. Vernberg WB, Scott GI, Strozier SH, Bemiss J, Daugomah JW (1996) The effects of urbanization on human and ecosystem health. In: Vernberg FJ, Vernberg WB, Siewicki TS (eds) Sustainable development in the southeastern coastal zone. University of South Carolina Press, Columbia, pp 221–239
103. WEF (2024) Nature positive: guidelines for the transition in cities, Insight Report. In collaboration with Oliver Wyman. https://www3.weforum.org/docs/WEF_Nature_Positive_Guidelines_for_the_Transition_in_Cities_2024.pdf
104. WHO (2012) Health indicators of sustainable cities in the context of the Rio+20 UN conference on sustainable development. WHO, Geneva
105. WHO (2017) Urban green spaces: a brief for action. WHO Regional Office for Europe, Bonn. https://www.who.int/europe/publications/i/item/9789289052498
106. WHO (2021) Urban health. https://www.who.int/news-room/fact-sheets/detail/urban-health
107. Wilkinson C, Macfarlane AH, Hikuroa DCH, McConchie C, Payne M, Holmes H, Mohi R, Hughes MW (2022) Landscape change as a platform for environmental and social healing. K¯otuitui N.Z. J Soc Sci Online 17:352–377
108. Wilson EO (1984) Biophilia: the human bond with other species. Harvard University Press, Cambridge, CA
109. Wolch JR, Byrne J, Newell JP (2014) Urban green space, public health, and environmental justice: the challenge of making cities 'just green enough'. Landsc Urban Plan 125:234–244
110. Wright G, Van der Heijden K, Burt G, Bradfield R, Cairns G (2008) Scenario planning interventions in organizations: an analysis of the causes of success and failure. Futures 40:218–236
111. Yan W, Roggema R, Middleton L (2017) Planning and design in Kesennuma: remember, reconnect, reform. In: Roggema R, Yan W (eds) Tsunami and Fukushima disaster: design for reconstruction. Springer, Dordrecht, Heidelberg, London, pp 95–112
112. Yu K (2021) The sponge city: planning, design and political design. In: Pelsmakers S, Newman N (eds) Architecture and the climate emergency—everything needs to change. RIBA Publishing, London, pp 47–55

113. Zhou W, Wang J, Cadenasso ML (2017) Effects of the spatial configuration of trees on urban heat mitigation: a comparative study. Remote Sens Environ 195:1–12
114. Zipperer Wayne C, Northrop R, Andreu M (2020) Urban development and environmental degradation. Oxford research Encyclopedia of environmental. Science. https://oxfordre.com/environmentalscience/view/10.1093/acrefore/9780199389414.001.0001/acrefore-9780199389414-e-97

Climate-Responsive Coastal Design: Mangrove-Based Urban Resilience in Port Vila, Vanuatu

Ali Cheshmehzangi, Ayyoob Sharifi, Abolghasem Azhdari, Bao-jie He, Zaheer Allam, Jyotir Nisha, and Ali Rad Yousefnia

1 Introduction: Coastal Urban Vulnerability and Nature-Based Design

Some of the planet's most environmentally dynamic and climate-exposed landscapes are found in coastal communities around the South Pacific [31]. Numerous urban communities in the area are situated along low-lying islands, lagoon margins, and narrow coastal plains, where land is scarce, and environmental conditions are extremely vulnerable to climate change [28]. Sea level rise, more tropical cyclones, coastal flooding, and ecological degradation are all putting further strain on these cities [3, 20]. The vulnerability of coastal urban systems in the South Pacific has

A. Cheshmehzangi (✉) · A. Sharifi · A. Azhdari · B. He · Z. Allam · J. Nisha · A. R. Yousefnia
School of Architecture, Design and Planning, The University of Queensland, Brisbane, QLD, Australia
e-mail: A.chesh@uq.edu.au

A. Cheshmehzangi
College of Architecture and Urban Planning, Tongji University, Shanghai, China

A. Sharifi
The IDEC Institute & Network for Education and Research on Peace and Sustainability (NERPS), Hiroshima University, Hiroshima, Japan

B. He
Centre for Climate-Resilient and Low-Carbon Cities, School of Architecture and Urban Planning, Key Laboratory of New Technology for Construction of Cities in Mountain Area, Ministry of Education, Chongqing University, Chongqing, China

State Key Laboratory of Subtropical Building and Urban Science, South China University of Technology, Guangzhou, Guangdong, China

CMA Key Open Laboratory of Transforming Climate Resources to Economy, Chongqing, China

Z. Allam
Institute for Sustainable Futures, University of Technology Sydney, Sydney, NSW, Australia

A. Cheshmehzangi et al. (eds.), *Nature-based and Climate Responsive Design for Urban Systems*, Urban Sustainability, https://doi.org/10.1007/978-981-92-2070-0_9

become a major concern for people, legislators, and planners alike as climate hazards increase.

Most climate adaptation studies focus on inequalities at the global scale, even though climate change often exacerbates existing social and spatial inequalities within countries [50], this is especially important in small island developing states, whose unique vulnerability has been repeatedly noted by scientific evaluations. Sea levels in the tropical Pacific have risen at rates faster than the global average in several places over the past few decades, according to the Intergovernmental Panel on Climate Change (IPCC). This has led to increased flood risks for low-lying settlements, coastal erosion, and saline intrusion into freshwater sources [20]. The bulk of the population lives in coastal areas in many Pacific Island states, such as Vanuatu, Fiji, Kiribati, and the Solomon Islands, along with important infrastructure, commercial hubs, and cultural landmarks [47]. As a result, urban areas like Port Vila, Suva, and Honiara must manage fast urban growth while adjusting to growing climatic challenges.

The vulnerability of coastal urban areas has been further highlighted by storm disasters [3]. Although the South Pacific climate system frequently experiences tropical cyclones, their severity has increased in recent decades. Cyclone Pam, a Category 5 hurricane that hit Vanuatu in March 2015, severely damaged Port Vila, the country's capital, and caused extensive damage throughout the archipelago [18]. The cyclone caused damage to dwellings, transportation networks, ecosystems, and coastal infrastructure, underscoring the fragility of urban communities constructed near shorelines without sufficient protection. The necessity of enhancing climate resilience in coastal urban development has been emphasised by similar experiences around the region.

Seawalls, revetments, and breakwaters are examples of man-made structures that have been used in traditional solutions to coastal dangers. Although these hard infrastructure solutions might offer instant protection in certain areas, they can have unforeseen environmental effects and are frequently costly to build and maintain [39]. For instance, seawalls can increase erosion in nearby areas, interfere with normal sediment transport processes, and harm coastal ecosystems like coral reefs and mangroves [39]. Furthermore, the dynamic climatic circumstances of tropical coastlines, where shifting sediments, tide patterns, and ecological processes impact the stability of coastal landscapes, might make inflexible engineering structures unsuitable.

Nature-based solutions (NbS) for coastal resilience have received more attention in recent years. Strategies that employ or restore natural ecosystems to address societal issues, including environmental degradation [26], disaster risk reduction, and climate adaptation, while also improving biodiversity and human well-being, are referred to as "nature-based solutions". These solutions are defined by the International Union for Conservation of Nature (IUCN) as measures that effectively and adaptively solve societal concerns by protecting, managing, and restoring natural or modified ecosystems [9, 21]. Mangrove restoration, coral reef conservation, wetland rehabilitation, and the creation of "living shorelines" that incorporate natural processes into coastal protection methods are some examples of these solutions in coastal contexts [23].

In tropical locations, mangrove forests in particular are important for coastal resilience. Mangrove resources and their management in the South Pacific have been the subject of sustained scholarly and policy attention for several decades. As early as 1981, Graham Baines highlighted the wide range of tangible benefits provided by mangrove ecosystems, identifying them as a source of structural timber, canoe logs, firewood, dyes, tannins, medicinal compounds, and diverse plant and animal-based food resources [4]. Beyond these material contributions, Baines also emphasised the critical intangible functions of mangroves, including their role in mitigating coastal erosion, protecting settlements from storm surges, supporting aquatic biodiversity, and trapping sediments derived from upstream soil erosion. Equally important is their cultural and social significance, as mangrove landscapes are deeply embedded in the livelihoods, traditions, and identities of Pacific Island communities. Mangroves stabilise sediments, lessen wave energy, and prevent erosion along shorelines by acting as natural buffers. Mangrove forests can significantly lessen the effects of storm surges and coastal floods by reducing wave heights and decreasing water flows, according to studies [29]. Mangroves offer crucial ecological services such as water filtration, fisheries habitat, and carbon sequestration in addition to their protective role. Mangroves are becoming more widely acknowledged as important "blue carbon" ecosystems [2, 52] that support attempts to mitigate climate change and reduce carbon emissions due to their ability to store carbon in both biomass and soils.

Coastal habitats have historically been managed through customary systems and local and traditional ecological knowledges and practices in the South Pacific region, which has a long history of community-based environmental care [17, 22]. Mangroves, reefs, and coastal forests are more than just natural resources in many island civilisations; they are essential components of social life, cultural identity, and food security [7, 27]. Therefore, acknowledging local and traditional knowledge systems and practices, as well as community involvement, is just as important as ecological restoration when incorporating nature-based techniques into urban development. In situations where local livelihoods are directly linked to coastal resources, community engagement is essential to the effective implementation of ecosystem-based adaptation measures.

As a result, nature-based coastal design has become a viable strategy for boosting resilience in metropolitan areas exposed to climate change. This method views ecosystems as active elements of urban systems that may carry out protective and regulatory tasks, as opposed to viewing nature as apart from urban infrastructure. Cities may create adaptable policies that address social needs and environmental change by combining ecological restoration with urban planning and citizen involvement.

Adopting nature-based coastal design offers South Pacific towns a chance to transition from reactive catastrophe management to more integrated and proactive resilience planning [8, 51]. Coastal ecosystem restoration and protection can simultaneously boost biodiversity, lower climate risks, and sustain local livelihoods. Developing coastal systems that complement natural processes rather than work against them will become a top goal for sustainable urban development in the area as urban populations continue to rise and climate pressures worsen.

Against this broader regional backdrop, this chapter focuses on Port Vila, Vanuatu, as a case through which to examine the role of mangrove-based coastal protection in climate-responsive urban design. It argues that mangrove restoration should be understood not only as an ecological intervention but also as a form of urban infrastructure with environmental, social, and planning implications. In doing so, the chapter uses the Port Vila case to explore how nature-based coastal design can support more resilient urban systems in Pacific Island contexts.

1.1 Nature-Based Coastal Infrastructure and Climate Adaptation

An increasingly significant strategy for climate adaptation in coastal areas across the globe is nature-based coastal infrastructure. This strategy acknowledges the protective potential of natural ecosystems and incorporates ecological processes into coastal planning and infrastructure design rather than depending exclusively on built structures like seawalls and breakwaters [30]. Ecosystems including mangroves, coral reefs, seagrass beds, salt marshes, and wetlands already serve as natural barriers in many coastal habitats, controlling wave energy, stabilising shorelines, and promoting biodiversity. These ecosystems may carry out many of the tasks often associated with hard infrastructure while providing additional ecological and socioeconomic advantages when they are successfully preserved or restored.

The idea of nature-based coastal protection is closely linked to the broader framework of ecosystem-based adaptation (EbA). The utilisation of biodiversity and ecosystem services as part of a comprehensive plan to assist communities in adjusting to the negative effects of climate change is known as "ecosystem-based adaptation" [42]. The goal of EbA initiatives in coastal environments is to make natural systems more resilient so that they can continue to protect coastal communities from environmental threats [43]. In tropical and subtropical areas, where ecosystems like mangroves and coral reefs are essential to preserving coastal stability, this strategy is especially pertinent.

One of the most well-known types of natural coastal infrastructure is mangrove forests. These intertidal ecosystems can be found in many tropical places, including the South Pacific, along protected river mouths, estuaries, and beaches. By stabilising coastal soils, retaining sediments, and reducing wave energy, mangroves serve a variety of protective purposes [13]. Mangrove trees can considerably lower the height and intensity of waves during storm events, reducing coastal erosion and flooding, according to research [29]. Their thick root systems produce organic barriers that improve silt deposition and reduce water flow, both of which support the integrity of the shoreline.

Mangroves provide a variety of ecosystem services that are vital to nearby people in addition to protecting the coast. In many islands and coastal civilisations, mangrove forests support coastal fisheries, serve as fish and crustacean nursery habitats, and enhance food security. Additionally, they are crucial for carbon storage. Mangrove habitats are known to be extremely effective carbon sinks, storing significant amounts of carbon in sediments and vegetation—a process known as "blue carbon" [1]. Thus, mangrove forest conservation and restoration support both climate adaptation and mitigation initiatives.

Another essential element of nature-based coastal defence is coral reef systems. As intricate underwater formations, coral reefs serve as natural breakwaters by lowering the energy of approaching waves before they reach coastlines. According to studies, coral reefs are among the best natural coastal defence systems since they can disperse up to 97% of wave energy [15]. For island nations and coastal towns that depend on reef systems as their first line of defence against storm surges and coastal floods, this protective role is especially crucial.

Additionally, coral reefs provide substantial ecological and financial benefits. They support fisheries that are essential to local livelihoods and offer habitat for a variety of marine creatures. Reef ecosystems are also essential to the tourism and recreational industries of many coastal economies. However, climate-related stressors such as ocean warming, coral bleaching, and ocean acidification are posing a growing threat to coral reefs globally. Reef ecosystems' capacity to defend coastlines diminishes as they deteriorate, underscoring the significance of conservation and restoration initiatives within more comprehensive climate adaptation plans.

Salt marshes and coastal wetlands both support flood control and shoreline protection. These ecosystems serve as organic barriers that trap sediments, absorb wave energy, and lessen the effects of storm surges. Over time, wetland vegetation can assist in elevating coastal areas by slowing water velocity and encouraging sediment deposition. Wetlands also serve as vital habitats for fish populations, bird species, and other types of biodiversity in many areas. Their importance in coastal urban systems is further increased by their ability to filter pollutants and enhance water quality.

In many coastal planning situations, the idea of "living shorelines" has evolved as a useful implementation of nature-based coastal infrastructure. To stabilise coasts while preserving or improving natural coastal processes, living shorelines employ natural materials and ecological restoration approaches. In order to develop adaptive coastal systems that change with the environment, these initiatives may incorporate vegetation planting, oyster reef restoration, and thoughtfully planned shoreline grading. Living shorelines provide more flexible long-term resilience than stiff seawalls because they can adapt to shifting sea levels and sediment patterns.

When opposed to conventional engineering methods, nature-based coastal infrastructure also offers social and economic benefits. Hard infrastructure projects can have minimal ecological advantages, large building costs, and continuous maintenance needs. Ecosystem-based solutions, on the other hand, can produce several co-benefits. For instance, restored mangrove forests may concurrently boost tourism, strengthen community resilience, support fisheries, and improve water quality [44].

Nature-based techniques are especially appealing for areas with limited financial and technical resources for large-scale engineering projects because of these multipurpose advantages.

Notwithstanding these benefits, the implementation of nature-based coastal infrastructure necessitates meticulous design, sustained administration, and robust governance frameworks. For ecosystem restoration efforts to be sustainable and successful, local environmental factors, hydrological processes, and community requirements must be considered [33, 38]. The effectiveness of these programmes frequently depends on community involvement, particularly in places where coastal resources are essential to local livelihoods. The preservation and restoration of coastal ecosystems are significantly influenced by traditional land management practices and ecological knowledge in many South Pacific locales.

As climate change continues to intensify pressures on coastal environments, integrating nature-based infrastructure into urban planning and coastal management is becoming increasingly important. By strengthening natural defence systems and lowering dependency on inflexible coastal barriers, nature-based techniques can supplement current infrastructure rather than completely replace built solutions. Coastal towns can create adaptation strategies that promote environmental sustainability and human well-being by preserving and restoring ecosystems, including mangroves, coral reefs, wetlands, and living shorelines.

2 Climate Pressures and Coastal Urban Systems in the South Pacific

The South Pacific's urban areas are among the world's most climate-exposed [8, 11, 41]. The bulk of the region's cities and towns are situated around beaches, estuaries, and low-lying islands where access to transportation routes, marine resources, and arable land has historically encouraged economic activity and population. These communities are now at the forefront of the effects of climate change due to their geographic advantage. Infrastructure, housing, and ecosystems that sustain urban life are increasingly at risk from sea-level rise, coastal erosion, storm surges, and extreme weather.

Rising sea levels are one of the most urgent issues affecting coastal communities in the South Pacific. Sea levels in a number of Pacific regions have risen more quickly than the world average in recent decades, according to observations made throughout the region [20]. Particularly in regions where urbanisation has already changed natural coastal processes, even small rises in sea level can greatly increase the frequency of coastal floods and hasten shoreline erosion. Numerous island capitals, such as Honiara in the Solomon Islands, Suva in Fiji, and Port Vila in Vanuatu, are situated on small coastal plains with little space for inland growth [46, 53]. The spatial limitations of these cities increase the risks related to coastal hazards as sea levels rise.

In several South Pacific cities, coastal flooding is also becoming more common. Storm surges, high tides, and periods of intense rainfall can all work together to overwhelm drainage systems and flood low-lying areas [10]. These concerns are especially severe in cities where the rise of informal settlements along sensitive shorelines has been fuelled by rapid population growth. Because they frequently lack durable dwelling structures, protective barriers, and proper drainage infrastructure, informal settlements are more vulnerable to environmental hazards such as flooding.

The resilience of coastal systems has been further undermined by environmental deterioration in addition to direct climatic impacts. Urbanisation, land reclamation, pollution, and unsustainable resource use have all contributed to the degradation of mangrove forests, coral reefs, and coastal wetlands, which were formerly natural protective buffers. The natural ability of coastlines to absorb wave energy, settle sediments, and control water flows is diminished when these ecosystems disappear. These natural protective systems have frequently been replaced by man-made protective systems like constructed structures and hardened surfaces due to coastal urbanisation, which may not offer the same long-term resilience.

The significance of ecosystem-based methods to climate adaptation has been increasingly highlighted by regional governments and international organisations in recognition of these difficulties. A regional policy, adopted in 2016, the Framework for Resilient Development in the Pacific (FRDP) encourages integrated approaches that integrate sustainable development, catastrophe risk reduction, and climate change adaptation [36]. As part of more comprehensive resilience planning for Pacific Island states, the framework promotes the preservation and repair of natural ecosystems for low-carbon growth and climate-resilient urban development.

In a similar vein, the region's national adaptation plans and climate policies have started to include nature-based approaches to urban development and coastal management. As part of their climate resilience agendas, nations including Fiji, Vanuatu, and Samoa have started initiatives centred on watershed protection, coral reef conservation, and mangrove restoration [12, 48]. These programmes enhance local livelihoods and biodiversity conservation while enhancing the protective roles of ecosystems (See Table 1).

In many Pacific communities, ecosystem stewardship has long been ingrained in traditional governing structures at the local level. The health of coastal ecosystems is still largely dependent on traditional management techniques, such as community-based conservation programmes and locally managed marine reserves [5]. These methods provide important institutional frameworks and knowledge that can support official policy initiatives to mitigate climate threats.

The ability to integrate biological processes into infrastructure construction and urban design will become increasingly crucial to the resilience of coastal urban systems in the South Pacific as climate pressures intensify. A strategy to improve coastal protection while maintaining the natural and cultural resources that characterise island populations is through ecosystem-based adaptation. By combining scientific knowledge, policy innovation, and local stewardship, Pacific cities can develop adaptive strategies that respond to both environmental change and the evolving needs of their urban populations. These wider regional pressures are clearly reflected in

Table 1 Key climate pressures, urban vulnerabilities, and ecosystem-based resilience strategies in South Pacific coastal cities

Dimension	Key Drivers	Urban Impacts	Examples in the South Pacific	Implications for Urban Resilience
Geographic Exposure	Coastal settlement patterns, low-lying islands, and proximity to marine resources	High vulnerability of infrastructure, housing, and transport networks to coastal hazards	Honiara (Solomon Islands), Suva (Fiji), Port Vila (Vanuatu) are located on narrow coastal plains	Requires spatial planning strategies that integrate coastal buffers, managed retreat options, and resilient infrastructure
Sea-Level Rise	Accelerated regional sea-level increase observed across Pacific basins	Increased coastal flooding frequency, shoreline erosion, and saline intrusion into freshwater systems	Pacific monitoring shows sea levels rising faster than the global average in parts of the region	Strengthening coastal planning regulations, elevation standards, and ecosystem restoration becomes critical
Extreme Weather & Flooding	Cyclones, storm surges, high tides, and intense rainfall events	Urban drainage failure, flooding of low-lying neighbourhoods, and infrastructure disruption	Coastal flooding events are reported in cities such as Suva and Port Vila during storm events	Hybrid drainage systems combining engineered and ecological solutions are increasingly necessary
Informal Urban Expansion	Rapid population growth and limited land availability	Informal settlements in vulnerable coastal zones lacking drainage, durable housing, or protection barriers	Informal coastal settlements around Honiara and Suva	Inclusive planning and climate-resilient housing policies are essential to reduce social vulnerability
Ecosystem Degradation	Mangrove clearing, coral reef damage, land reclamation, and pollution	Loss of natural buffers that absorb wave energy and stabilise sediments	Mangrove loss across many Pacific coastal towns	Ecosystem restoration becomes a cost-effective form of coastal infrastructure
Nature-Based Adaptation Policies	Regional and national climate frameworks promoting ecosystem-based adaptation	Integration of ecological systems into planning, coastal management, and disaster risk reduction	Framework for Resilient Development in the Pacific (FRDP); national adaptation plans in Fiji and Vanuatu	Policy alignment enables ecological infrastructure to become part of formal urban resilience strategies

(continued)

Table 1 (continued)

Dimension	Key Drivers	Urban Impacts	Examples in the South Pacific	Implications for Urban Resilience
Traditional Governance and Local Stewardship	Community-based conservation practices and locally managed marine areas	Protection of fisheries, reefs, and mangrove systems supporting livelihoods	Community marine reserves in Fiji and Samoa	Indigenous knowledge systems provide governance models for ecosystem-based resilience
Future Urban Adaptation Pathways	Integration of science, policy innovation, and local knowledge	More adaptive urban systems capable of responding to climate uncertainty	Growing initiatives in mangrove restoration and watershed protection across Pacific states	Urban resilience will increasingly depend on combining ecological restoration with infrastructure planning

Port Vila, which provides a useful case for examining how ecosystem-based coastal protection can be integrated into urban resilience planning.

3 Methodological Framework

A strategy that integrates environmental analysis, geographic exploration, and socio-institutional perspectives is necessary to understand climate-responsive coastal design. Coastal resilience is a complicated interplay between physical systems, governance structures, and community activities rather than just an ecological or engineering problem. In order to assess the contribution of nature-based coastal infrastructure to enhancing urban resilience, this chapter's methodological framework employs a mixed analytical approach that incorporates policy review, geographical analysis, and qualitative observations.

The case study technique that underpins the research design enables a thorough analysis of climate adaptation tactics in a particular metropolitan setting. Because case study research makes it possible to examine intricate socio-ecological systems where several variables interact at once, it is frequently employed in studies of urban and environmental planning. It is feasible to evaluate how climatic threats, environmental circumstances, governance structures, and community actions influence adaptation outcomes in practice by concentrating on a specific geographic area.

The study's data sources come from a number of complementary streams. The physical features of the coastal system and the sensitivity of metropolitan areas to climatic hazards are first understood via the use of environmental and geographical data. These include coastline conditions, coastal land-use patterns, satellite photography, and accessible geographic information system (GIS) datasets. The study can investigate how ecological infrastructure contributes to coastal protection and flood mitigation by using spatial analysis to identify locations where natural ecosystems, such as mangroves and wetlands, meet with urban development (See Fig. 1). In practice, this spatial analysis is used to identify where coastal ecosystems, shoreline conditions, and urban development intersect in Port Vila, rather than to produce a detailed quantitative hazard model.

The institutional and governance framework pertaining to coastal management and climate adaptation is then examined through a policy and planning assessment. This entails examining regional resilience policies, coastal planning frameworks, and national climate adaptation initiatives that are pertinent to the South Pacific region. The analysis takes into account how policy frameworks include ecosystem-based methods and how these frameworks either facilitate or hinder the development of nature-based coastal infrastructure.

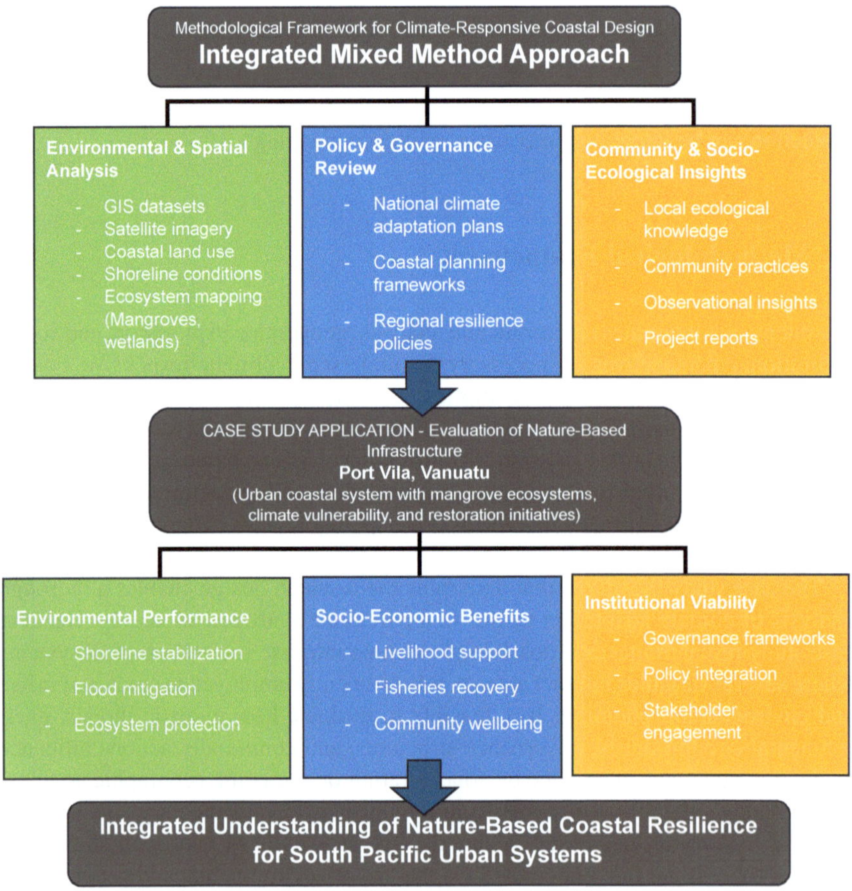

Fig. 1 Methodological framework integrating spatial analysis, policy review, and socio-ecological observations to evaluate nature-based coastal resilience in Port Vila, Vanuatu

Third, community viewpoints and regional environmental practices are recorded through qualitative observations and secondary documentation. In the South Pacific, where traditional ecological knowledge and customary governance systems frequently impact the management of coastal ecosystems, community-based observations are especially crucial. Contextualising how nature-based tactics function in typical urban and community settings is made easier by observational insights, which are bolstered by project reports and local documentation where available.

Port Vila, Vanuatu, was selected as the case study for this chapter based on a number of factors. These include the urban coastal zone's substantial mangrove ecosystems, its vulnerability to climate-related coastal hazards, and the existence of ongoing ecosystem restoration projects meant to increase coastal resilience. Additionally, the city is a classic example of a quickly expanding capital on a Pacific Island where climatic vulnerability and urban growth pressures collide.

Three important factors are taken into consideration while evaluating nature-based resilience measures in the case study: institutional viability, socioeconomic advantages, and environmental performance. The capacity of ecosystems like mangroves to maintain shorelines and lessen exposure to coastal risks is considered when evaluating environmental performance. Socioeconomic advantages look at how fisheries, livelihoods, and community well-being are impacted by ecological restoration. Institutional viability evaluates how community involvement, regulatory frameworks, and governance structures support long-term ecosystem management.

The chapter aims to offer a thorough understanding of how nature-based coastal infrastructure might support climate adaptation in coastal urban systems throughout the South Pacific using this integrated analytical approach.

3.1 Case Study Context: Port Vila, Vanuatu

An essential case study for researching climate-responsive coastal design in the South Pacific is Vanuatu's capital, Port Vila. One of the most important urban and commercial hubs in the area, the city is situated on the island of Efate near the protected waters of Mele Bay [8]. Because of its coastal location, which has historically facilitated trade, fishing, tourism, and transportation, Port Vila serves as a significant entry point that links Vanuatu to regional and global networks [35, 37]. In addition, the city's location puts it in a highly climate-exposed area where coastal hazards represent a growing threat to community livelihoods and urban development.

Coastal lowlands, mangrove-lined shorelines, coral reef systems, and inland highlands combine to form Port Vila's physical landscape. Numerous urban districts are concentrated along the harbour and along the shore, where homes, businesses, and infrastructure have grown up close to the water's edge. This pattern of development reflects both the geographic constraints of the island environment and the historical importance of maritime access. However, it also increases vulnerability since coastal communities are more vulnerable to storm surges, sea-level rise, and coastal floods.

Port Vila has had consistent urbanisation and population increase over the last few decades. The city draws people from other islands looking for work, education, and services because it is Vanuatu's main administrative and commercial hub. The development of peri-urban and informal communities along low-lying and coastal areas has been facilitated by this population rise. These communities have frequently grown up in areas particularly vulnerable to the effects of storms, erosion, and flooding. These communities' susceptibility to environmental risks is further increased by inadequate housing and infrastructure.

These preexisting vulnerabilities are getting worse due to climate change. Coastal urban settings are facing issues due to coastal erosion and saltwater intrusion into freshwater systems caused by rising sea levels around the Pacific. Several areas of Efate Island have experienced shoreline erosion, which has an impact on developed infrastructure and coastal ecosystems. Coastlines are now more vulnerable to erosion and wave action in places where mangrove forests have been destroyed or damaged.

The dangers that Port Vila's coastal inhabitants face are exemplified by extreme weather events. Storm Pam, a Category 5 tropical storm that hit Vanuatu in March 2015 [14], was one of the most important recent events. Infrastructure, housing, agriculture, and coastal habitats were all severely damaged nationwide by Cyclone Pam. Strong winds and storm surges at Port Vila caused damage to buildings, interrupted vital services, and brought attention to how susceptible urban systems are to extreme weather occurrences [34]. Additionally, the cyclone seriously damaged coastal ecosystems and flora, which typically help to safeguard the shoreline.

In Vanuatu's urban planning and environmental management policies, the aftermath of Cyclone Pam brought disaster risk reduction and climate adaptation back into focus. The necessity of bolstering biological systems and physical infrastructure that can lower climate threats was brought to light by recovery efforts [34]. More attention has been paid to the protective roles of coastal ecosystems, including coral reefs and mangroves. Mangrove habitats around Port Vila, particularly those around river mouths and estuarine zones, have come to be recognised as important natural resources that may stabilise shorelines [24] and mitigate the effects of storms.

Therefore, Port Vila offers a timely and pertinent case study for investigating how nature-based coastal protection might be integrated into urban systems. The city is a typical example of a Pacific Island capital that is expanding quickly, and where climatic risks, environmental change, and urban growth come together. Mangrove ecosystems and current environmental restoration projects, however, present opportunities to investigate how NbS can support climate resilience.

This case study helps highlight the more general difficulties faced by coastal urban systems in the South Pacific by looking at Port Vila's geographic setting, development trends, and climate exposure. The city's experiences show how vulnerable island communities are to the effects of climate change as well as how ecosystem-based solutions may help improve long-term urban resilience.

3.2 Case Study Analysis: Mangrove Restoration and Coastal Resilience

The importance of mangrove ecosystems in the Port Vila area for coastal resilience in Vanuatu's urban setting is becoming increasingly acknowledged. In the past, mangroves created vast intertidal forests along protected shorelines, estuaries, and river mouths around Mele Bay. These ecosystems supported fisheries, biodiversity, and coastal livelihoods while acting as natural barriers that shielded coastlines from erosion and wave energy. But in many parts of Port Vila, mangrove coverage has decreased due to decades of urban growth, land reclamation, and infrastructure development, undermining the natural defence mechanisms that formerly shielded the city's coastline.

As part of more comprehensive ecosystem-based adaptation strategies, recent efforts have focused on restoring mangrove ecosystems in response to increasing climate risks and environmental degradation. Partnerships with regional and international organisations, community-based projects, and national environmental programmes all contribute to these restoration efforts. The objective is to increase the resilience of urban communities exposed to coastal risks while restoring damaged coastal ecosystems.

Mangrove restoration is crucial for maintaining shorelines and enhancing the health of coastal ecosystems from an environmental standpoint. Mangrove root systems gradually construct and stabilise coastal soils by capturing sediments delivered by river discharge and tidal flows. By retaining sediment, a natural barrier against wave movement is created and shoreline erosion is lessened. Restored mangrove areas along sheltered coasts and estuary zones in Port Vila have shown promise in slowing coastal erosion and lessening the susceptibility of nearby land regions to tidal flooding and storm surges.

Mangroves also aid in the restoration of coastal ecosystems and the improvement of water quality. Their intricate root systems filter nutrients, contaminants, and sediments from water that flows into coastal lagoons and reefs from metropolitan areas. In addition to promoting healthier marine ecosystems, this filtering method reduces the risk of sediment buildup, which can harm seagrass beds and coral reefs. As mangrove ecosystems recover, they help restore ecological connectivity between terrestrial and marine environments, which is essential for maintaining coastal biodiversity.

Another significant result of mangrove restoration in the Port Vila coastal zone is the recovery of biodiversity. Mangrove forests sustain local fisheries by serving as nursery grounds for a variety of fish species, crabs, and molluscs [19]. Additionally, these ecosystems give birds and other creatures somewhere to nest and feed. Mangrove habitat regeneration helps these biological networks gradually recover as restoration efforts grow [6]. For coastal communities whose livelihoods and food security depend on marine environments, bolstering local fishing resources is especially crucial.

Mangrove restoration projects in Port Vila have produced significant social and community results in addition to environmental advantages (See Table 2).

Mangrove planting, monitoring, and conservation management are among the many restoration programmes that engage local communities [45]. In situations where coastal resources are intimately connected to local livelihoods and cultural customs, community involvement is essential to the long-term sustainability of ecosystem restoration initiatives. Traditional ecological knowledge is frequently incorporated into community-based conservation programmes, which can help choose suitable restoration sites and species.

Programmes for mangrove restoration now heavily rely on environmental education and public awareness [16]. In an effort to raise awareness of the value of coastal ecosystems, local schools, community organisations, and civil society organisations have taken part in mangrove planting projects. These programmes promote community management of coastal areas and increase public awareness of the protective roles of mangroves.

Mangrove restoration supports a more comprehensive strategy for managing coastal risks from the standpoint of urban resilience. Ecosystem-based solutions offer adaptive protection that can change as the environment does, as opposed to solely depending on man-made coastal defences [29]. Over time, mangrove forests develop and spread, enabling them to adapt to changing coastal dynamics and rising

Table 2 Analytical summary of mangrove restoration contributions to coastal resilience in Port Vila, Vanuatu

Dimension	Key Mechanism	Observed Outcomes in Port Vila	Implications for Coastal Resilience
Shoreline stabilisation	Mangrove root systems trap sediments and reduce wave energy	Reduced shoreline erosion and improved coastal stability in sheltered lagoons	Natural buffering of storm surges and tidal flooding
Ecosystem recovery	Mangrove forests restore ecological connectivity between land and marine systems	Improved water quality and habitat recovery in lagoon environments	Healthier coastal ecosystems and long-term environmental resilience
Biodiversity and fisheries	Mangroves provide nursery habitats for fish, crabs, and molluscs	Gradual recovery of fisheries resources and marine biodiversity	Strengthened food security and livelihood support
Community engagement	Community-based restoration, monitoring, and education initiatives	Increased public awareness and local participation in mangrove conservation	Stronger stewardship and long-term sustainability of restoration programmes
Adaptive coastal protection	Nature-based solutions evolve with changing coastal dynamics	Mangrove forests expand and adjust to sediment and tidal changes	Flexible and cost-effective protection compared to rigid coastal engineering

sea levels. They are especially useful in settings where long-term climatic uncertainties call for adaptable and sustainable protection policies because of their potential for adaptation.

However, the effort of incorporating mangrove restoration into urban development is still in its early stages. Coastal ecosystem protection may face difficulties due to conflicting development agendas, infrastructure growth, and land-use pressures. Therefore, maintaining the resilience benefits of mangrove ecosystems requires incorporating them into coastal management policies and urban development frameworks.

Overall, Port Vila's experience shows how mangrove restoration can aid in both urban climate resilience and environmental recovery. These programmes demonstrate how Nature-based coastal infrastructure can support traditional adaptation measures by improving biodiversity, stabilising shorelines, and boosting community involvement. Ecosystem-based strategies like mangrove restoration present a viable solution to create more resilient and sustainable urban coastal systems as coastal communities around the South Pacific continue to face growing climate shocks and stresses.

4 Discussion: Integrating Nature-Based Coastal Design into Urban Systems

This section reflects on the broader implications of the Port Vila case for urban planning and climate adaptation. The integration of nature-based solutions into urban infrastructure development, community involvement, and governance issues is all covered. Insights into the broader role of nature-based coastal design in urban systems vulnerable to climate change can be gained from the Port Vila scenario [8]. Although ecosystem protection and mangrove restoration are frequently presented as environmental projects, their effects go far beyond ecological preservation. These ecosystems serve as useful infrastructure that can concurrently assist catastrophe risk reduction, climate adaptation, and sustainable urban growth in the South Pacific's fast urbanising coastal cities [51]. The Port Vila experience serves as an example of how incorporating biological systems into urban planning can help coastal communities become more adaptable and resilient.

The connection between patterns of urban expansion and the destruction of coastal ecosystems is one of the case's major ramifications [40]. Due to the scarcity of land and the economic significance of marine access, urban growth has traditionally occurred along coasts in many Pacific Island communities. Mangrove clearing, wetland restoration, and shoreline modification to make way for housing, transportation infrastructure, and commercial activity have all frequently resulted from this pattern. Short-term urban growth may be aided by these measures, but they also eliminate natural coastal buffers that stabilise sediments, control wave energy, and lower the risk of flooding.

The Port Vila study shows that the environmental effects of such growth patterns can be somewhat mitigated by restoring and safeguarding these ecosystems. Mangrove restoration initiatives have demonstrated how natural coastal infrastructure can enhance shoreline stability while also promoting community livelihoods and biodiversity [25, 49]. These restoration initiatives, however, draw attention to a significant planning issue: if ecosystem protection is to maintain its long-term advantages, it must be integrated into larger land-use planning frameworks. Restored habitats are nonetheless susceptible to future pressures from urban expansion if they are not formally acknowledged in zoning laws, development restrictions, and coastal management plans.

Therefore, the success of nature-based coastal adaptation solutions is largely dependent on governance. Coastal management tasks are divided among several entities in several South Pacific contexts, including national government agencies, local governments, customary landowners, and community organisations. Coordinating efforts for coastal preservation or ecosystem restoration may be difficult due to this disjointed governance structure. Clear institutional duties, supportive regulatory frameworks, and cooperation among sectors like environmental management, urban planning, fisheries, and disaster risk reduction are necessary for the successful integration of nature-based solutions into urban infrastructure development.

Policy alignment is also critical. Ecosystem-based adaptation is becoming more widely acknowledged as a crucial tactic for mitigating climate risks in national climate adaptation plans and resilience frameworks around the Pacific. It is still difficult to translate these governmental pledges into local planning procedures, though. If nature-based solutions are to become a key element of climate-resilient urban systems, building codes, coastal infrastructure investments, and urban development rules must align with ecosystem protection goals. The Port Vila scenario indicates that improved collaboration between municipal planning procedures and national adaptation programmes could greatly increase the efficacy of ecosystem-based coastal protection.

Another essential component of nature-based coastal resilience is community involvement. Coastal ecosystems are intimately linked to traditional resource management techniques and land tenure structures in many Pacific nations. Local governance mechanisms that have developed over many generations are frequently used to manage fishing areas, mangroves, and reefs cooperatively. Involving local populations in planning, execution, and long-term stewardship activities is crucial, as seen by the success of mangrove restoration projects in Port Vila.

Participation from the community enhances socio-cultural resilience as well as ecological resilience. When locals actively participate in restoration initiatives, they become more committed to preserving coastal habitats and become more conscious of environmental hazards. A sense of ownership and shared accountability for preserving natural defences can be fostered by this involvement. Additionally, community-based conservation efforts frequently combine scientific methods with traditional ecological knowledge to provide adaptation solutions that are both culturally sensitive and regionally appropriate.

The conceptual framing of ecosystems as infrastructure inside urban systems is another crucial topic for consideration. Stormwater networks, drainage systems, and seawalls are examples of designed solutions that have been the main emphasis of traditional urban infrastructure development. Although these systems are still crucial for controlling urban settings, they frequently need strict and expensive interventions to mitigate climate threats. A complementary viewpoint is provided by nature-based coastal design, where ecosystems carry out infrastructural tasks while preserving natural processes.

For instance, mangroves can be thought of as living infrastructure that develops and changes over time. Mangrove ecosystems may react dynamically to environmental changes like shifting sediment flows and rising sea levels, in contrast to static concrete constructions. Because of their ability to adapt, nature-based solutions are especially useful in the setting of climate uncertainty, where long-term environmental conditions are not always predictable.

But incorporating ecological infrastructure into urban development also necessitates overcoming real-world obstacles. The amount of space available for ecosystem restoration in urban coastal locations may be constrained by land-use competition, development constraints, and short-term economic considerations. Therefore, planners, policy makers and legislators must think about how to distribute urban land in a way that strikes a balance between the necessity for development and the preservation of ecosystems. Within growing urban landscapes, strategic coastal zoning, ecological buffer zones, and integrated watershed management techniques may offer ways to preserve space for nature-based infrastructure.

Ultimately, the Port Vila case shows that nature-based coastal design should be seen as an essential part of climate-resilient urban systems rather than as a stand-alone environmental intervention. Incorporating ecosystems into urban planning frameworks can improve biodiversity, bolster community livelihoods, and promote coastal protection all at once. Future coastal urban settings' resilience will be greatly influenced by cities' capacity to coordinate governance, ecological restoration, and citizen involvement as climate pressures increase throughout the South Pacific.

5 Conclusions and Lessons for the South Pacific Region

The Port Vila study demonstrates the increasing significance of nature-based coastal design as a workable and efficient means of enhancing urban resilience in island regions exposed to climate change. Mangrove restoration and ecosystem-based coastal protection programmes can provide several advantages that go beyond environmental preservation, as this chapter has shown. These ecosystems directly support community livelihoods, biodiversity recovery, storm surge reduction, and coastline stabilisation. Therefore, restoring and safeguarding mangrove systems can be crucial in enhancing the adaptability of urban coastal landscapes in quickly urbanising coastal areas like Port Vila.

Nature-based coastal infrastructure should be acknowledged as an essential part of urban systems rather than as a peripheral environmental intervention, according to one of the analysis's main results. Wetlands, coral reefs, and mangrove forests serve as living infrastructure that naturally guard against climatic threats. Their ecological activities enhance sediment stabilisation, lessen wave energy, and control coastal dynamics. These systems can greatly increase long-term resilience while lowering reliance on costly constructed coastal defences when incorporated into coastal management and urban development frameworks.

The need for governance and policy congruence in facilitating ecosystem-based adaptation is another crucial lesson. The Port Vila experience serves as an example of how, although national climate policies are beginning to acknowledge the significance of nature-based solutions, their implementation at the urban level necessitates greater integration with infrastructure development, land-use planning, and environmental management systems. In order to safeguard restored ecosystems against future development pressures and guarantee their long-term sustainability, clear institutional coordination and supportive regulatory frameworks are crucial.

Another important element in the success of mangrove restoration projects is community involvement. Coastal ecosystems are intricately linked to local livelihoods, customary land management, and cultural practices in many South Pacific contexts [32]. Involving local populations in planting, monitoring, and stewardship activities through restoration initiatives can increase environmental awareness and promote a sense of collective responsibility for the preservation of ecosystems. Climate adaptation tactics can be made more successful and culturally relevant by combining scientific planning techniques with traditional ecological knowledge.

For other South Pacific coastal cities and island communities that deal with comparable environmental and development challenges, the lessons learned from Port Vila are especially pertinent. Similar features, such as coastal settlement patterns, restricted land availability, and a heavy reliance on marine ecosystems, are shared by many Pacific urban centres. Because of these factors, nature-based coastal protection is a suitable and scalable approach to boosting resilience throughout the area.

In the future, a balanced blend of biological restoration, urban design, and community involvement will be necessary for climate-responsive coastal resilience in the South Pacific. In addition to their environmental importance, ecosystems like mangroves should be preserved because they help shield urban areas from the effects of climate change. Pacific cities may create adaptable coastal systems that promote human well-being and environmental sustainability in an increasingly uncertain climate future by coordinating ecosystem protection with urban growth policies.

References

1. Alongi DM (2014) Carbon sequestration in mangrove forests. Carbon Manage 5(3):285–296
2. Alongi DM (2020) Global significance of mangrove blue carbon in climate change mitigation. Science 2(3):67
3. Australian Bureau of Meteorology, & CSIRO (2022) State of the climate 2022. Commonwealth of Australia. Available online from: https://www.csiro.au/state-of-the-climate
4. Baines G (1981) Mangrove resources and their management in the South Pacific, Part of the 'South Pacific regional environment programme', SPREP/Topic Review 5, Available online from: https://www.spc.int/DigitalLibrary/Doc/FAME/Reports/Baines_81_MangroveResources_Management_Pacific.pdf
5. Beger M, Harborne AR, Dacles TP, Solandt JL, Ledesma GL (2004) A framework of lessons learned from community-based marine reserves and its effectiveness in guiding a new coastal management initiative in the Philippines. Environ Manage 34(6):786–801
6. Bosire JO, Dahdouh-Guebas F, Walton M, Crona BI, Lewis RR III, Field C, Kairo JG, Koedam N (2008) Functionality of restored mangroves: a review. Aquatic Bot 89(2):251–259
7. Campbell JR (2015) Development, global change and traditional food security in Pacific Island countries. Reg Environ Change 15(7):1313–1324
8. Cheshmehzangi A (2026) From data to action: using multi-hazard maps to shape Vanuatu's disaster strategy. In: He B, Cheshmehzangi A, Mohaghegh L, Mittal M (eds) Designing for a changing climate: a science-based path to practical actions. Springer, Singapore
9. Cohen-Shacham E, Walters G, Janzen C, Maginnis S (2016) Nature-based solutions to address global societal challenges. IUCN: Gland, Switzerland 97:2036
10. Connell J (2018) Effects of climate change on settlements and infrastructure relevant to the Pacific islands. Pacific Marine Climate Change Report Card, UK Government, pp 159–176
11. Consalo K (2022) Vulnerable populations: climate change and extreme weather threats facing urban communities. Chi Kent J Env't Energy L 11:1
12. Devi J, Holland EA (2024) Building resilience in a climate crisis: best practices for mangrove restoration along the Coral Coast, Fiji. Pac Dyn J Interdisc Res 8(1):503–521
13. Duke NC, Mackenzie J, Wood A (2012) A revision of mangrove plants of the Solomon Islands, Vanuatu, Fiji, Tonga and Samoa. Centre for Tropical Water & Aquatic Ecosystem Research (TropWATER) Publication 12:13
14. Durutalo AL, Hernawan B, Nanau GL, Van Trease H (2017) Melanesia in review: issues and events. The Contemporary Pacific pp 341–373
15. Ferrario F, Beck MW, Storlazzi CD, Micheli F, Shepard C, Airoldi L (2014) The effectiveness of coral reefs for coastal hazard risk reduction and adaptation. Nat Commun 5:3794
16. Friess DA, Thompson BS, Brown B, Amir AA, Cameron C, Koldewey HJ, Sasmito SD, Sidik F (2016) Policy challenges and approaches for the conservation of mangrove forests in Southeast Asia. Conserv Biol 30(5):933–949. https://doi.org/10.1111/cobi.12784
17. Govan H, Aalbersberg W, Tawake A, Parks J (2008) Locally managed marine areas: a guide to supporting community-based adaptive management. Published by: The Locally Managed Marine Area Network
18. Government of Vanuatu (2015) Vanuatu post-disaster needs assessment: tropical Cyclone Pam, March 2015. Government of Vanuatu
19. Hutchison J, Spalding M, Zu Ermgassen P (2014) The role of mangroves in fisheries enhancement. The Nature Conservancy and Wetlands International 54:434
20. IPCC (2021) Climate change 2021: the physical science basis. Cambridge University Press, Cambridge
21. IUCN (2016) Nature-based solutions to address global societal challenges. International Union for Conservation of Nature. Available from: https://portals.iucn.org/library/sites/library/files/documents/2016-036.pdf
22. Johannes RE (2002) The renaissance of community-based marine resource management in Oceania. Annual Rev Ecol Syst 33:317–340. https://doi.org/10.1146/annurev.ecolsys.33.010802.150524

23. Jordan P, Fröhle P (2022) Bridging the gap between coastal engineering and nature conservation? a review of coastal ecosystems as nature-based solutions for coastal protection. J Coastal Conserv 26(2):4. https://doi.org/10.1007/s11852-021-00848-x
24. Komugabe-Dixson AF, de Ville NS, Trundle A, McEvoy D (2019) Environmental change, urbanisation, and socio-ecological resilience in the Pacific: community narratives from Port Vila, Vanuatu. Ecosystem Serv 39:100973
25. Lovelock CE, Barbier E, Duarte CM (2022) Tackling the mangrove restoration challenge. PLOS Biol 20(10):e3001836
26. Mackey B, Ware D, Nalau J, Buckwell A, Sahin O, Fleming C, Smart J, Connolly R, Hallgren W (2017) Vanuatu Ecosystem and Socio-economic Resilience Analysis and Mapping (ESRAM). Published by Secretariat of the Pacific Regional Environment Programme (SPREP), Apia, Samoa
27. Macusi ED, Katikiro RE, Deepananda KA, Jimenez LA, Conte AR, Fadli N (2011) Human induced degradation of coastal resources in Asia Pacific and implications on management and food security. J Nat Stud 9(10):13–28
28. Magnan AK, Oppenheimer M, Garschagen M, Buchanan MK, Duvat VKE, Forbes DL, Ford JD, Lambert E, Petzold J, Renaud FG, Sebesvari Z, van de Wal RSW, Hinkel J, Pörtner HO (2022) Sea level rise risks and societal adaptation benefits in low-lying coastal areas. Sci Rep 12(1):10677. https://doi.org/10.1038/s41598-022-14303-w
29. McIvor A, Tonneijk FH (2014) Mangroves for coastal defence. Guidelines for coastal managers & policy makers. Published by Wetlands International and The Nature Conservancy, p 42
30. Morris RL, Konlechner TM, Ghisalberti M, Swearer SE (2018) From grey to green: efficacy of eco-engineering solutions for nature-based coastal defence. Global Change Biol 24(5):1827–1842. https://doi.org/10.1111/gcb.14063
31. Nurse LA, McLean RF, Agard J, Briguglio LP, Duvat-Magnan V, Pelesikoti N, Tompkins E, Webb A (2014) Small islands. Climate change 2014: Impacts, adaptation, and vulnerability. Part B: Regional aspects. Contribution of working group II to the fifth assessment report of the intergovernmental panel on climate change, p 1613
32. Pacific Community (SPC) (2016) Framework for Resilient Development in the Pacific (FRDP). Developed and published by Pacific Islands Forum Secretariat, & partners. Available from: https://gem.spc.int/projects/frdp
33. Pander J, Geist J (2013) Ecological indicators for stream restoration success. Ecol Indic 30:106–118
34. Potangaroa R (2015) Review of the IFRC shelter cluster coordination services in response to TC Pam in Vanuatu. 70 pages Review report, Available online from: https://unhcr-sheltercluster-static.s3.eu-west-1.amazonaws.com/public/docs/vanuatu_tc_pam_sc_review_2015.pdf
35. Poustie MS, Frantzeskaki N, Brown RR (2016) A transition scenario for leapfrogging to a sustainable urban water future in Port Vila, Vanuatu. Technol Forecast Social Change 105:129–139
36. Savage A, Huber C, Bambrick H (2020) Integration of climate change adaptation, disaster risk reduction and development in Vanuatu: a systems perspective. Handbook of governance in Small states. Routledge, New York and London, pp 151–166
37. Sitko P, Toma W, Johnson O (2020) Using a systems approach to better understand urban resilience in Port Vila, Vanuatu. Urbanisation at risk in the Pacific and Asia. Routledge, New York and London, pp 89–106
38. Suding KN (2011) Toward an era of restoration in ecology: successes, failures, and opportunities ahead. Annual Rev Ecol Evol Syst 42(1):465–487
39. Temmerman S, Meire P, Bouma T, Herman P, Ysebaert T, De Vriend H (2013) Ecosystem-based coastal defence in the face of global change. Nature 504(7478):79–83. https://doi.org/10.1038/nature12859
40. Todd PA, Heery EC, Loke LH, Thurstan RH, Kotze DJ, Swan C (2019) Towards an urban marine ecology: characterizing the drivers, patterns and processes of marine ecosystems in coastal cities. Oikos 128(9):1215–1242

41. Trundle A, Barth B, McEvoy D (2019) Leveraging endogenous climate resilience: urban adaptation in Pacific Small island developing states. Environ Urban 31(1):53–74
42. UNEP (2019) Ecosystem-based adaptation for climate resilience. United Nations Environment Programme. Available online from: https://www.unep.org/topics/climate-action/adaptation/ecosystem-based-adaptation
43. Vasseur L (2021) How ecosystem-based adaptation to climate change can help coastal communities through a participatory approach. Sustainability 13(4):2344. https://doi.org/10.3390/su13042344
44. Veitayaki J, Waqalevu V, Varea R, Rollings N (2017) Mangroves in small island development states in the Pacific: an overview of a highly important and seriously threatened resource. Participatory mangrove management in a changing climate: perspectives from the Asia-Pacific, pp 303–327
45. Walters BB, Rönnbäck P, Kovacs JM, Crona B, Hussain SA, Badola R, Primavera JH, Barbier E, Dahdouh-Guebas F (2008) Ethnobiology, socio-economics and management of mangrove forests: a review. Aquatic Bot 89(2):220–236. https://doi.org/10.1016/j.aquabot.2008.02.009
46. Weaver S, Renwick J, Rastandeh A, Maxwell D, Gual CMA, Loubser D, Livesey CC, Komugabe-Dixson A, Jackson B, Archie KM, Zari MP (2017). Ecosystems, climate change and development scenarios, Vanuatu
47. Weir T, Dovey L, Orcherton D (2017) Social and cultural issues raised by climate change in Pacific Island countries: an overview. Regional Environ Change 17(4):1017–1028. https://doi.org/10.1007/s10113-016-1012-5
48. Weir T, Dovey L, Okrupa M (2025) Ecosystem-based adaptation in the Pacific islands. In Handbook of nature-based solutions to mitigation and adaptation to climate change. Springer Nature Switzerland, Cham, pp 1139–1162
49. Wilson AMW, Forsyth C (2018) Restoring near-shore marine ecosystems to enhance climate security for island ocean states: aligning international processes and local practices. Mar Policy 93:284–294
50. Zahnow R, Rad Yousefnia A, Hassankhani M, Cheshmehzangi A (2025) Climate change inequalities: a systematic review of disparities in access to mitigation and adaptation measures. Environ Science Policy 165:104021. https://doi.org/10.1016/j.envsci.2025.104021
51. Zari MP, Kiddle GL, Blaschke P, Gawler S, Loubser D (2019) Utilising nature-based solutions to increase resilience in Pacific ocean cities. Ecosyst Serv 38:100968
52. Zeng Y, Friess DA, Sarira TV, Siman K, Koh LP (2021) Global potential and limits of mangrove blue carbon for climate change mitigation. Current Biol 31(8):1737–1743
53. Zhu B, Bai Y, He X, Chen X, Li T, Gong F (2021) Long-term changes in the land–ocean ecological environment in small island countries in the South Pacific: a Fiji vision. Remote Sensing 13(18):3740

Cities Must Learn from Nature or Fail with It: From Adaptation to Transformation by Reimagining Urban Systems with Nature

Ali Cheshmehzangi, Jian Zuo, Ayyoob Sharifi, Rongpeng Zhang, Abbas Ziafati Bafarasat, and Jie Zhao

1 From Evidence to Imperative: What the Cases Reveal

The case studies presented in this volume collectively go much beyond demonstrating the applicability of nature-based and climate-responsive design, exposing more profound change in the way urban systems need to be perceived and organized. Urban resilience is now attained through alignment with ecological processes rather than control in a variety of geographical locales, climates, and socioeconomic realities. This is demonstrated by experience; it is not a theoretical claim.

The crucial function that water plays as a structuring element in urban areas is among the most startling findings found throughout the chapters. Water highlights the

A. Cheshmehzangi (✉) · A. Sharifi
School of Architecture, Design and Planning (ADP), The University of Queensland, Brisbane, QLD, Australia
e-mail: A.Chesh@uq.edu.au

J. Zuo
School of Civil Engineering and Construction Management, Adelaide University, Adelaide, SA, Australia

A. Sharifi
The IDEC Institute and Network for Education and Research on Peace and Sustainability (NERPS), Hiroshima University, Hiroshima, Japan

A. Cheshmehzangi · R. Zhang
School of Architecture and Planning, Hunan University, Changsha, China

A. Ziafati Bafarasat
School of the Built Environment, Oxford Brookes University, Oxford, UK

J. Zhao
The Weitzman School of Design, The University of Pennsylvania, Philadelphia, PA, USA

P3Global, Dubai, UAE

A. Cheshmehzangi et al. (eds.), *Nature-based and Climate Responsive Design for Urban Systems*, Urban Sustainability, https://doi.org/10.1007/978-981-92-2070-0_10

shortcomings of traditional systems as well as the possibilities of alternative strategies, whether in densely populated areas, small coastal communities, or quickly urbanizing populations. Water scarcity, erosion, and flooding are not isolated problems; rather, they are systemic results of design choices that have given impermeability, quick discharge, and spatial efficiency precedence over retention, infiltration, and ecological balance. The case studies show how water may concurrently change urban design, improve environmental performance, and increase social value when it is viewed as an asset rather than a danger.

The transition from single-function infrastructure to multifunctional landscapes is closely related to this. Conventional systems are usually made to carry out a single function, such as stabilizing land, protecting against coastal surges, or draining water. On the other hand, the projects discussed in this book demonstrate how nature-based methods function simultaneously in several dimensions. For example, a blue-green corridor serves as a microclimatic regulator, an ecological habitat, a public area, and a drainage system. Similarly, vegetative ground cover techniques improve the standard of daily urban surroundings while also aiding in soil health, temperature regulation, and water management. This stacking of functions is essential to how nature-based systems produce resilience; it is not an accident.

The significance of context-driven solutions is another important realization. Standardized models or universal templates are not used in the examples that are discussed. Rather, local factors such as resource availability, government arrangements, cultural traditions, and climate variability shape them. Adaptive, gradual, and resource-efficient solutions are frequently found in cities with severe climate stresses and limited institutional and financial resources. This calls into question the widely held belief that significant investment or advanced technology are prerequisites for innovation. Conversely, some of the best tactics are those that take advantage of current circumstances and make use of local resources and expertise to deliver results that are appropriate for the setting.

The stories also highlight the shortcomings of traditional, engineering-led solutions. Although crucial in some situations, grey infrastructure frequently follows strict guidelines and is created for particular thresholds. These systems are being pushed harder and harder as climatic unpredictability increases, which can result in expensive and disruptive breakdowns. Nature-based systems, on the other hand, show a level of resilience and flexibility that enables them to react to shifting circumstances without catastrophically failing. This suggests a shift towards hybrid approaches that include biological processes into infrastructure design rather than a total replacement of constructed systems.

Crucially, the data also emphasizes resilience's social component. The best results come from integrating nature-based solutions into everyday community life, such as designing public areas to accommodate water, incorporating local practices into environmental management, and implementing interventions that improve not only environmental performance but also liveability, accessibility, and social cohesion. This supports the notion that resilience is a spatial and social state rather than just a technical result.

Taken together, these observations shift the conversation from demonstration to necessity. The viability of nature-based and climate-responsive techniques is no longer in doubt. Whether urban systems can afford to function without them is the question. The body of evidence indicates that they are unable to. These examples show a definite path rather than a collection of isolated successes: cities need to change from systems built to oppose nature to ones that can function with, through, and as part of ecological processes.

1.1 Beyond Isolated Interventions: Towards Systemic Integration

The case studies show that despite being effective, nature-based and climate-responsive techniques have significant shortcomings in present practices. Many interventions are still project-based, fragmented, and not fully integrated into larger urban systems. Decentralized drainage systems, retention landscapes, and green corridors are frequently put into practice as stand-alone or prototype projects that are isolated from more extensive infrastructure networks and planning frameworks. When these treatments are not included into a cohesive, system-wide strategy, their influence is limited even when they produce localized benefits.

This fragmentation is a reflection of larger structural problems with the planning, financing, and governance of cities rather than just a design problem. Water, transportation, land use, and the environment are examples of sectoral silos that are commonly used to organize urban systems. Each sector has its own set of objectives, rules, and investment cycles. In contrast, nature-based solutions naturally transcend these constraints. They necessitate alignment between short-term implementation and long-term strategic planning, as well as collaboration between disciplines and institutions. Their potential to support resilience at scale remains underutilized in the absence of such integration.

Therefore, switching from projects to systems thinking is necessary to get beyond discrete actions. This entails incorporating nature-based methods into planning tools, legal frameworks, and infrastructure plans so that they are viewed as essential elements of urban development rather than as optional extras. For instance, water-sensitive design principles must govern public space design rules, zoning laws, and citywide drainage planning in addition to individual developments. In a similar vein, ecological networks need to be designed as cohesive systems rather than as isolated green areas.

The temporal aspect of implementation presents another difficulty. Many nature-based treatments change over time, necessitating ongoing supervision, management, and adjustment. Conventional project cycles, on the other hand, are frequently brief and prioritize delivery above continuous performance. The efficacy of nature-based systems, which rely on ongoing ecological processes and iterative adjustment, is limited by this mismatch. In order to address this, new governance models that emphasize care over completion are needed, acknowledging that resilience is a continuous state rather than a predetermined result.

Ultimately, the transition towards systemic integration is not about linearly scaling up individual solutions. It involves rethinking how urban systems are designed and run in order to incorporate natural processes at all scales and sectors. Despite their value, isolated initiatives are insufficient in the face of escalating climatic pressures, as demonstrated by the research given in this book. The transition to integrated, multifunctional, and adaptive urban systems, where nature is a defining framework for urban resilience rather than an add-on, must be intentional and well-coordinated.

2 Scaling Nature-Based Urbanism: From Site to Region

The question of how nature-based solutions may be effectively scaled, rather than whether they are effective, is a recurrent issue in both the case studies and more general urban practice. Scaling is sometimes confused with replication, which is the expansion of successful enterprises into multiple sites. The research in this book, however, indicates that this strategy is inadequate. Nature-based urbanism grows through system alignment across institutional structures, ecological processes, and geographical layers rather than just repetition.

Interventions like permeable surfaces, vegetal ground cover, decentralized drainage, and blue-green public areas show definite advantages at the site level. They enhance biodiversity, lessen runoff, improve local microclimates, and promote social wellness. However, how they are linked outside of their immediate surroundings determines how much of an impact they have overall. Urban hydrology cannot be substantially changed by a single green space or retention area unless it is a component of a networked system that functions throughout districts and neighbourhoods.

This highlights how crucial inter-scalar design thinking is. Urban systems are enmeshed in broader environmental systems that span administrative and geographic scales rather than being limited to distinct borders. Ecological corridors span the boundaries between urban and rural areas, water flows across catchments, and regional processes influence coastal dynamics. Understanding cities as nested systems, where interventions at one scale affect and depend on conditions at another, is therefore necessary for scaling nature-based approaches.

In practice, this requires a shift from isolated site-based solutions to coordinated spatial strategies. In order to connect parks, streetscapes, waterways, and open spaces into a cohesive system, blue-green infrastructure must be designed as continuous networks. In a similar vein, catchment-level planning that takes into account both upstream and downstream interactions should incorporate water-sensitive urban design. Mangroves, wetlands, and natural buffers that function at the landscape scale must be included in coastal resilience initiatives that go beyond city limits.

Planning and governance structures are another crucial aspect of scaling. Even well-designed programs have a limited reach without institutional support. Zoning laws, development restrictions, infrastructure planning, and investment strategies must all incorporate nature-based concepts to scale. In order to guarantee that interventions are coordinated rather than dispersed, it also requires coordination across sectors, including planning, engineering, environmental management, and community development. Scaling is therefore both a governance and a design challenge.

Crucially, contextual variation must also be taken into consideration while scaling. The case studies show how local characteristics, such as geography, climate, socioeconomic considerations, and cultural norms, have a significant impact on effective solutions. Therefore, uniform models cannot be relied upon for scaling. Rather, it necessitates the creation of flexible frameworks and transferable ideas that may be adjusted to various situations. This strategy maintains implementation flexibility while allowing for consistency in aim, such as improving infiltration, reestablishing ecological connectedness, or lowering heat stress.

Lastly, scaling nature-based urbanism necessitates a change in the metrics used to assess success. Evaluation must consider system-wide results, such as cumulative environmental benefits, enhanced resilience across scales, and long-term adaptability, rather than concentrating just on the success of individual projects. To capture the dynamic nature of ecological systems, new metrics, monitoring strategies, and feedback mechanisms are needed.

Scaling is essentially about reorganizing urban systems such that nature-based practices become ingrained, interconnected, and self-reinforcing across scales rather than doing more of the same. The entire potential of nature-based urbanism in tackling the challenges of the climate age can only be realized through this systematic alignment.

3 Reframing Urban Systems: Nature as Core Infrastructure

The information gathered throughout this book suggests that nature has to be acknowledged and handled as essential urban infrastructure. This is a structural change rather than a rhetorical one. It questions the established hierarchy that treats biotic systems as supplemental, ornamental, or residual and constructed systems, roads, pipes, barriers, and grids as primary. This structure is becoming more and more unworkable in the climatic age. Ecological processes must be integrated as fundamental elements of urban performance to ensure urban resilience, which cannot be achieved through constructed systems alone.

By framing nature as infrastructure, one is redefining the role of infrastructure. Conventional infrastructure, which usually operates within predetermined parameters and provides one or a small number of functions, is built for efficiency, predictability, and control. Water is removed by a drainage system, waves are blocked by a seawall, and mobility is made easier by a transportation network. Although these systems work well in stable environments, they are frequently inflexible, resource-intensive, and prone to failure when conditions surpass design thresholds. These constraints are revealed by climate variability, as systems are increasingly forced to operate beyond their designed capacity.

Ecological systems, on the other hand, function by ongoing adjustment, redundancy, and adaptation. Mangroves attenuate wave energy while renewing over time; wetlands expand and compress in response to water levels; and vegetated surfaces collect, filter, and release moisture. These systems lessen the possibility of an abrupt and catastrophic failure by distributing and modulating risk rather than completely eliminating it. Their ability to absorb shocks, recover, and change is what defines their performance rather than just resistance. This presents an infrastructural logic that is fundamentally different and more in line with the uncertainty and fluctuation that characterize the climatic period.

But designating nature as essential infrastructure does not mean that engineered systems must be completely replaced. Instead, it advocates for the creation of hybrid infrastructures, which combine artificial and ecological elements. In these systems, natural systems offer flexibility, adaptation, and co-benefits, whereas grey infrastructure offers stability and tailored performance. For instance, rather than depending on a single line of defence, a flood control system can integrate built channels with floodplains, wetlands, and permeable landscapes to distribute water across several pathways. This integration lessens the strain on any one component while increasing the overall resilience of the system.

This reframing has important ramifications for how infrastructure is evaluated, funded, and maintained. Conventional investment models favour designed solutions that can be standardized and measured, giving priority to infrastructure with well-defined outputs and quantifiable returns. In contrast, ecological infrastructure offers a variety of interrelated benefits that are more challenging to quantify under traditional cost-benefit frameworks, such as social value, water management, biodiversity support, and climate regulation. Therefore, broadening evaluation metrics to take into consideration long-term performance, co-benefits, and saved costs rather than just short-term efficiency is necessary to recognize nature as infrastructure.

To facilitate this transition, governance frameworks must also change. Urban growth and environmental management are often not well-coordinated, and infrastructure planning is often dispersed across sectors. Cross-sectoral alignment, the integration of planning, design, engineering, and ecological knowledge within cohesive frameworks, is necessary for integrating nature as infrastructure. Because ecological systems require continuous management, monitoring, and adaptation rather than one-time delivery, it also necessitates long-term stewardship approaches.

This reframing has significant spatial implications. Nature must be given room inside the urban fabric, not as marginal or leftover land, but as a key organizing element, if it is to serve as infrastructure. This could entail rethinking land-use distribution to accommodate water and plants as dynamic systems, restoring biological corridors, and rearranging urban design to prioritize open space networks. In high-density settings where land is highly contested, these modifications provide a challenge to traditional development logics. However, they are necessary for cities to continue operating in the face of shifting environmental conditions.

Finally, acknowledging nature as essential infrastructure is a step towards reinventing the function and efficiency of urban systems. It moves the emphasis from discrete functions to interconnected systems, from static design to adaptive processes, and from opposing external influences to interacting with them. Rather than simplifying the problems cities face, this reframing recognizes their complexity and offers a more adaptable, integrated, and resilient solution. This consequently lays the groundwork for urban systems that can both adapt to and survive the effects of climate change.

3.1 Policy, Governance, and Implementation Pathways

Reframing nature as essential infrastructure necessitates a systemic change in governance, policy, and implementation methods in addition to innovative design. Even the most successful nature-based approaches are limited to discrete projects in the absence of institutional coherence, making it impossible for them to have an impact on the overall operation of urban systems. Therefore, enabling frameworks that include nature-based approaches into the regular logic of planning and development are necessary to promote the shift towards integrated, ecological urbanization.

Policy integration is a crucial first step. To ensure that nature-based solutions are not viewed as optional elements but rather as essential elements of urban development, they must be included into statutory planning instruments, development restrictions, and infrastructural recommendations. This includes requiring blue-green infrastructure in new construction, integrating water-sensitive design concepts into zoning laws, and coordinating climate adaptation plans with land-use planning. By leading urban transformation instead of reacting to crises, such integration changes the role of policy from reactive to proactive.

The necessity of cross-sectoral governance is equally crucial. Water management, urban planning, environmental preservation, and community development are all areas where nature-based infrastructure functions. Therefore, institutional cooperation is necessary for effective implementation, dismantling the silos that frequently prevent coordinated action. New governance models that support integrated decision-making, shared responsibility, and long-term cooperation across agencies and stakeholders are needed for this.

To facilitate this shift, financing methods must also adapt. Conventional infrastructure investment sometimes ignores the long-term, multi-benefit worth of natural systems in favour of initiatives with obvious, short-term rewards. Funding models must take into consideration co-benefits including decreased flood risk, better public health, and increased ecosystem services in order to address this issue, acknowledging that these results support social and economic resilience. Scaling implementation may be aided by creative finance strategies like blended funding, public–private partnerships, and climate adaptation funds.

However, implementation is a continuous process. Monitoring, upkeep, and adaptive management are all part of the continuous stewardship needed for nature-based systems. As a result, the emphasis is shifted from project delivery to long-term performance, where success is determined by the system's capacity to change over time in addition to its original results. Thus, maintaining these systems requires building institutional capacity through training, knowledge sharing, and community involvement.

The road forward is ultimately determined by the convergence of governance, investment, and practice around a common goal: integrating nature into the fundamental operations of urban systems, rather than by discrete policies or initiatives. This alignment becomes not just desirable but also essential as cities deal with increasing climate stresses. The problem is now operational rather than intellectual.

4 The Next Urban Paradigm: Designing Cities that Coexist with Nature

The data in this book leads to a conclusion that is impossible to deny and even harder to soften: how fast and seriously cities learn to cooperate with nature will decide their future. Not as a catchphrase or a style of design, but as a fundamental idea that transforms the conception, management, and maintenance of urban systems. The ease of gradual adjustment has been eliminated by the climate era. The future is a time of systemic redefinition rather than gradual improvement.

Cities have functioned for decades under the presumption that environmental constraints could be managed by larger systems, stronger materials, and more accurate calculations. That is no longer a reasonable assumption. The scope and regularity of climatic impacts have shown a basic discrepancy between the behaviour of natural systems and urban planning. The question now is whether cities can afford to stick with their existing course rather than if they can continue on it.

This book offers a path rather than a single model: cities that move from resilience as resistance to resilience as adaptation, and finally towards co-evolution and regeneration. More than simply technological changes are needed for this changeover. It necessitates a re-evaluation of priorities, where land is allotted for water, vegetation, and ecological processes in addition to development; infrastructure is assessed not only for efficiency but also for its ability to adapt, absorb, and sustain; and design is viewed as a long-term interaction with dynamic systems rather than the provision of fixed solutions.

However, the way forward is neither clear-cut nor assured. It is influenced by conflicting interests, institutional inertia, regional disparities in capacity, and economic pressures. The difficulty lies not only in proving that nature-based strategies are effective, but also in integrating them into the frameworks that determine how cities develop and operate. This necessitates addressing challenging issues pertaining to government, funding, and societal values in addition to design.

In situations where land is limited and heavily commodified, can cities reallocate space to ecological systems? Before climate impacts surpass planning cycles, can policy frameworks advance quickly enough to incorporate nature-based approaches? Can models of infrastructure investment prioritize long-term ecological benefits over immediate financial gains? Can really integrated, cross-sectoral action be made possible by governance systems overcoming fragmentation? Most importantly, can urban societies change their expectations such that they accept interdependence with nature instead of trying to control it?

Although there are no easy solutions or answers to these concerns/questions, they do set the stage for discussions about the future of urban systems. Nonetheless, it is evident that the price of doing nothing or taking only limited action is increasing. As the effects of climate change worsen, cities that still rely on inflexible, single-function systems will experience more disruption, higher maintenance costs, and widening social disparities. Those that embrace ecological integration, on the other hand, will more successfully disperse, absorb, and adapt to risk rather than completely eliminate it.

This book describes a transition that is not optional. As evidenced by the initiatives, tactics, and practices discussed throughout the chapters, it is already in progress in fragmented and unequal ways. Moving from isolated instances to systemic change, from experimentation to mainstream practice, is the current challenge. Thus, cities must learn from nature, or they will fail with it.

Index

A. Cheshmehzangi et al. (eds.), *Nature-based and Climate Responsive Design for Urban Systems*, Urban Sustainability, https://doi.org/10.1007/978-981-92-2070-0

MIX
Papier aus verantwortungsvollen Quellen
Paper from responsible sources
FSC® C105338
FSC
www.fsc.org

If you have any concerns about our products,
you can contact us on
ProductSafety@springernature.com

In case Publisher is established outside the EU,
the EU authorized representative is:
Springer Nature Customer Service Center GmbH
Europaplatz 3, 69115 Heidelberg, Germany

Printed by Libri Plureos GmbH
in Hamburg, Germany